T0135918

Proceedings

of the

Beilstein Bozen Symposium

on

MOLECULAR ENGINEERING AND CONTROL

May 14th – 18th, 2012

Prien (Chiemsee), Germany

Edited by Martin G. Hicks and Carsten Kettner

**Beilstein-Institut**

Beilstein Bozen Symposium on Molecular Engineering and Control
May 14th – 18th, 2012, Prien (Chiemsee), Germany

BEILSTEIN-INSTITUT ZUR FÖRDERUNG DER CHEMISCHEN WISSENSCHAFTEN

Trakehner Str. 7 – 9
60487 Frankfurt
Germany

| **Telephone:** | +49 (0)69 7167 3211 | **E-Mail:** | info@beilstein-institut.de |
| **Fax:** | +49 (0)69 7167 3219 | **Web-Page:** | www.beilstein-institut.de |

IMPRESSUM

Molecular Engineering and Control, Martin G. Hicks and Carsten Kettner (Eds.), Proceedings of the Beilstein Bozen Symposium, May 14th – 18th 2012, Prien (Chiemsee), Germany.

Bibliographic information published by the *Deutsche Nationalbibliothek*. The *Deutsche Nationalbibliothek* lists this publication in the *Deutsche Nationalbibliografie*; detailed bibliographic data are available in the Internet at http://dnb.ddb.de.

ISBN 978-3-8325-3606-0

Layout by: Hübner Electronic Publishing GmbH
             Steinheimer Straße 22a
             65343 Eltville
             www.huebner-ep.de

Printed by: Logos Verlag Berlin GmbH
              Comeniushof, Gubener Str. 47
              10243 Berlin
              www.logos-verlag.de

Cover Illustration by: Bosse und Meinhard
                       Kaiserstraße 34
                       53113 Bonn
                       www.bosse-meinhard.de

PREFACE

The Beilstein Bozen Symposia address contemporary issues in the chemical and related sciences by employing an interdisciplinary approach. Scientists from a wide range of areas – often outside chemistry – are invited to present aspects of their work for discussion with the aim of not only to advance science, but also, to enhance interdisciplinary communication.

With the increasing understanding of molecular systems it is now possible to build materials and new systems with nano-scale precision through the control of the structure of matter at the atomic and molecular level. Since the forces that dominate the macroscopic world have either less relevance or different consequences at the nano-level, we must employ different paradigms when conceiving molecular-scale machines that will build, in turn, new types of materials and machines, etc. In this respect biological systems are the best proof of concept that this kind of technology already exists. Multicomponent systems such as ribosomes can be considered as molecular-scale machines that read RNA, decode the information, generate proteins and finally assist in the folding process to ensure the generation of a correct three-dimensional configuration. This newly created entity can carry out structural functions, catalytic activities in chemical processes, and even form a constituent part of further ribosomes for the construction of new molecular machines. Inspired by biological systems, researchers are beginning to mimic nature in the design of molecules and supramolecular systems but also in the modification of nature's own factories.

A key aim is to be able to routinely design molecules or systems with desired physicochemical or physiological properties.

For example, the manipulation and control of molecules on surfaces to bring about the functionalization of the surface or of the molecules themselves is important for a wide variety of applications. Accomplishing this requires not only expertise in synthesis but also in many other techniques such as imaging, lithography and computation. Many difficulties associated with being able to simultaneously understand and control assembly, recognition, transport and motion at the molecular and systems levels remain and need to be addressed by future research.

This symposium brought together experts from different disciplines to discuss, from their own points of view, the contemporary state and future perspectives including the following aspects of molecular engineering and control, i.e. molecular control of surfaces, manipulation of metabolic pathways and engineering of proteins and nucleotides, self-organization and molecular self-assembly, imaging, diagnostics and sensors, and artificial (biological) systems.

We would like to thank particularly the authors who provided us with written versions of the papers that they presented. Special thanks go to all those involved with the preparation and organization of the symposium, to the chairmen who piloted us successfully through the sessions and to the speakers and participants for their contribution in making this symposium a success.

Frankfurt/Main, December 2013

Martin G. Hicks
Carsten Kettner

CONTENTS

Page

Molecular Control of (Stem) Cell Fate

Dave Winkler

CSIRO Materials Science & Engineering, Clayton 3168, Australia and
Monash Institute of Pharmaceutical Sciences, Parkville 3052, Australia

E-Mail: Dave.Winkler@csiro.au

Received: 18th October 2012 / Published: 13th December 2013

Introduction

The notion of cell identity or phenotype has undergone a seismic shift over the past decade. Until then, cell biologists largely regarded terminally differentiated somatic (i. e. non-germ line) cells as deriving from more plastic progenitors via an essentially one-way route. Only recently was the question of reversibility of cell differentiation, a by-product of the inherent stochasticity and plasticity of cells, raised by researchers such as Roeder and Loeffler [1]. The explosion of research into stem cells over the past decade in particular has vindicated these early suggestions of mutability and plasticity of cell phenotypes. A recently as 2006, Yamanaka announced the startling discovery that somatic cells can be reprogrammed to pluripotency by a cocktail of transcription factors [2]. Subsequent research has shown that it may be possible to reprogram somatic cells of one type into those of a different type, such as reprogramming skin epithelial cells to neural cells. The idea that a cell's identity is better described as a probabilistic property than a fixed one is now becoming more widely accepted.

Although work to date on cell reprogramming and other forms of cellular transformation, such as directed differentiation of a pluripotent or multipotent progenitor to a terminally differentiated cells, has largely relied on genetic or viral modification of cells, there is a small but rapidly increasing interest in the role that specifically designed small organic molecules may have in cellular reprogramming. This chapter summarizes some of the progress towards small molecule control of cell fate, and chemically induced cell reprogramming. The ability to have fine control over cell identity and fate will clearly lead to major medical advances in tissue and organ regeneration and cancer, now increasingly thought to have aberrant stem cells as a significant cause.

I will briefly summarize the properties and potential uses of stem cells, describe the role of gene regulatory networks in controlling cell fate decision and providing the origin of cell plasticity or stochastic behaviour, discuss how aberrant stem cell programming can drive cells towards detrimental phenotypes, then summarize early progress in the use of small organic molecules to control the fate of cells and drive transitions between different cell phenotypes:

- Pluripotent to somatic
- Somatic to pluripotent
- Somatic to somatic
- Aberrant pluripotent to somatic or death

The chapter will finish with a brief perspective of the future for small molecule-induced cellular reprogramming.

PROPERTIES AND POTENTIAL USES OF STEM CELLS

Stem cells are cells with multiple differentiation options (Fig. 1). There are essentially three types: germ line stem cells, about which nothing further will be discussed; pluripotent stem cells (of which embryonic stem cells are a subset); and adult stem cells that have an essential role in maintaining bodily tissues that wear out, are damaged, or lost.

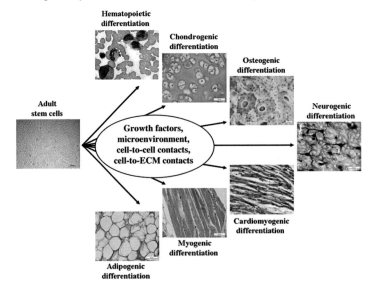

Figure 1. Multipotential capacity of embryonic stem or iPS cells to generate terminally differentiated tissues under the action of environmental factors. (Danišovič *et al.* (2012) *Exp. Biol. Med.* **237**:10 – 17)

Embryonic stem cells (ES cells) are transient pluripotent stem cells derived from the inner cell mass of the embryonic blastocyst. They possess two distinctive properties: pluripotency (can generate cells of almost any type): ability to replicate almost indefinitely.

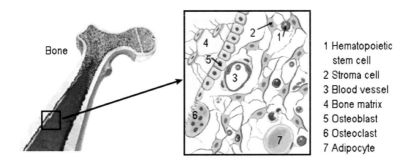

Bone

1 Hematopoietic
  stem cell
2 Stroma cell
3 Blood vessel
4 Bone matrix
5 Osteoblast
6 Osteoclast
7 Adipocyte

Figure 2. Illustration of HSC niche in the endosteal region of the bone marrow showing some of the niche components.

Adult stem cells are undifferentiated cells, found throughout the body, that divide to replenish dying cells and regenerate damaged tissues. They reside in very specific '*niches*' that control their fates (see figure 2 for an illustration of a haematopoietic stem cell niche). They possess two important properties: self-renewal (the ability to go through numerous cycles of cell division while still maintaining its undifferentiated state); and multipotency or multidifferentiative potential-the ability to generate progeny of specific but limited cell types.

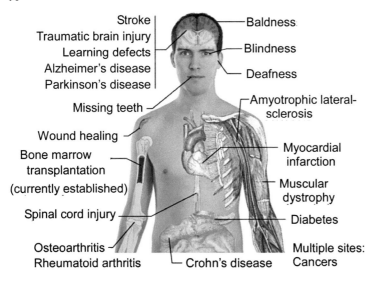

Stroke
Traumatic brain injury
Learning defects
Alzheimer's disease
Parkinson's disease
Missing teeth
Wound healing
Bone marrow
transplantation
(currently established)
Spinal cord injury
Osteoarthritis
Rheumatoid arthritis

Baldness
Blindness
Deafness
Amyotrophic lateral-
  sclerosis
Myocardial
infarction
Muscular
dystrophy
Diabetes
Crohn's disease
Multiple sites:
Cancers

Figure 3. A summary of potential stem cell therapies for major diseases.

Because pluripotent stem cells have the potential to become many types of somatic cell types if their differentiation can be controlled, they offer incredible potential as therapies to replace worn, diseased, or damaged parts of the body. Reprogrammed cells have the additional advantage of allowing the patient's own cells to be transformed into new tissue, largely overcoming any immune rejection that allografts often encounter. Potential uses for stem cell therapies are summarized in figure 3.

CELLS ARE STOCHASTIC OBJECTS, PLURIPOTENCY IS A PROBABILISTIC PROPERTY – CELL PLASTICITY

Cells and cell phenotypes were once thought to be fixed, and transitions between different cell types largely biologically or molecularly impossible because of epigenetic imprinting. Cell gene regulatory networks control phenotype, and cell state is now seen to be a potentially reversible, probabilistic state. Expressed phenotypes can be quite heterogeneous, regulatory trajectories can take multiple paths to the same endpoint, and cellular reversion or dedifferentiation is possible. Cell phenotypes are also heterogeneous, hinting at the underlying stochastic behaviour of gene expression, which is nonetheless still tightly regulated. This stochasticity is illustrated in figure 4 by the substantial fluctuations in expression of a key pluripotency transcription factor Nanog in mouse ES cells [3].

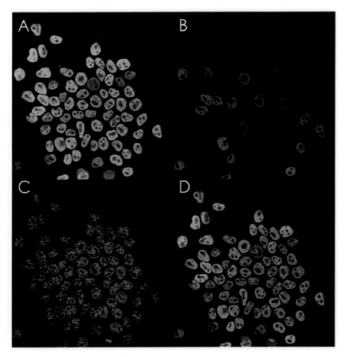

Figure 4. Immunofluorescence staining for **(A)** Oct4 and **(B)** Nanog, and **(C)** staining with DAPI; **(D)** an overlay of A-C. In mouse ES cells, Oct4 staining appears to be relatively homogeneous, whereas Nanog expression levels differ substantially within individual ES cells [3]. Roeder and Radtke (2009), *Development:* Image courtesy of Austin Smith.

Not only is expression of cell surface marker genes in stem cells stochastic and heterogeneous, but the trajectory of gene expression that is followed from one cell state to another can also be highly heterogeneous, as shown by Huang in elegant experiments summarized in figure 5 [4]. He drove neutrophils into differentiation using retinoic acid or DMSO. Both differentiation triggers resulted in the neutrophils differentiating into the same phenotypic cell but analysis of the gene expression profile of both processes showed that the trajectories were markedly different. This shows that cells can start at the same point but transit completely different regulatory gene expression programs before arriving at a common endpoint.

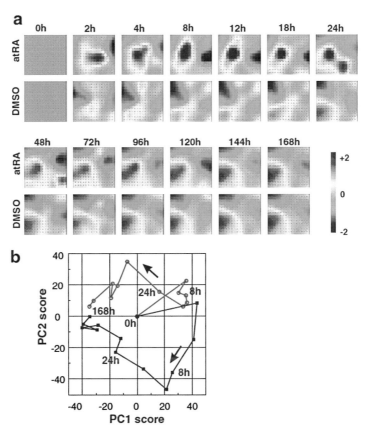

Figure 5. Comparison of the two gene expression trajectories during neutrophil differentiation. **(a)** The genes were clustered by a self-organizing map into 15 × 16 "miniclusters" with regard to their temporal profiles across both differentiation processes. Tile colors indicate the expression level of the cluster centroid; numbers on color bar: gene expression levels in SLR units. **(b)** Principal component analysis. Each point represents an individual expression profile S(t) within one of the two differentiation processes (red circles: RA; blue squares:DMSO) projected onto the first two principal components. Huang *et al.* (2005) PRL 94, 128701 [4]

These gene regulatory programs have been likened to mutable information networks where connections between nodes (genes) are made and broken depending on the presence and promoter/repressor binding of relevant transcription factors that control gene expression. Clearly, other control mechanisms involving microRNAs, changes to chromatin structure and epigenetic marks, and the presence of external cues such as growth factors, cytokines, adhesion molecules, cell-matrix and cell-cell mechanical and chemical cues can also influence the state and operation of gene regulatory networks (Fig. 6).

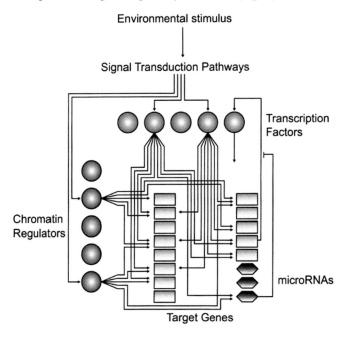

Figure 6. Representation of gene regulation as a control process. Marson Doctoral thesis MIT 2008

These networks are rich in regulatory loops, suggesting a complex system exhibiting a wide range of context-dependent dynamic behaviours. Such dynamic networks exhibit a set of stationary states called attractors (Fig. 7) that have been suggested by Kauffman to correspond to the observed number of different types of cells of the body [5]. Cell states and transitions between them can therefore be visualized as features on a gene regulatory surface or landscape, a term first coined by Waddington [6]. This simple but powerful description is relevant to the discussion of cellular reprogramming below.

Complex networks of gene interactions have a limited number of stationary attractor states. Kauffman hypothesized that these states correspond to stable cell phenotypes, and the region of states near an attractor that lead to that attractor is called the basin of attraction.

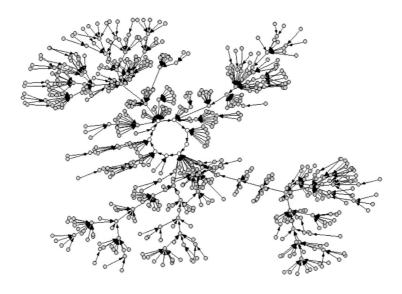

Figure 7. Attractors can be fixed (point) or cyclic. Kauffman (1969) *J. Theor. Biol.* **22:**437 – 467.

CELLULAR REPROGRAMMING AS NAVIGATION THROUGH A COMPLEX ATTRACTOR LANDSCAPE

In a complex cellular attractor landscape there might be many coexisting stationary attractors (here represented as local minima), each of which might be associated with a unique molecular signature. In this view, cellular reprogramming corresponds to guiding the cell through the landscape from one local minimum to another (shown by the dotted arrows) [7]. As there might be many distinct paths between minima (both direct and through intermediary minima), reprogramming from one cell type to another might be achieved through numerous different routes (Fig. 8).

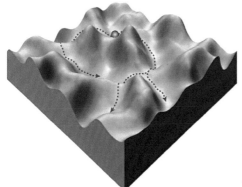

Figure 8. Waddington's epigenetic landscape. Macarthur *et al.* (2009) *Nat. Rev. Mol. Cell Biol.*

The valleys represent stable cell attractor (stationary) states generated by a hypothetical regulatory network. Depending on the particular configuration of the network (e. g. different parameter values, such as transcription or decay rates), a different number and/or different qualities of attractors are possible [3].

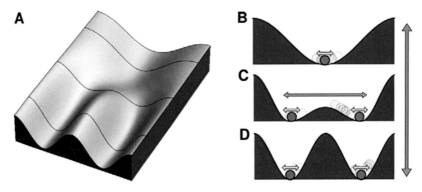

Figure 9. Cell fates or phenotypes as stable attractors (valleys) in regulatory landscape. Roeder and Radtke (2009) *Development.*

MODELLING STEM CELL FATE DECISIONS

Simplified models of key switching mechanisms in pluripotent cells such as the SONs (Sox-2, Oct4, Nanog) network that maintains pluripotency [7], or the GATA-1/PU.1 switch that controls HSC differentiation fate, can be modelled using a number of different mathematical methods. Nonlinear dynamical theory, agent-based modelling, Boolean networks, and machine learning methods are common mathematical modelling techniques that have been applied to modelling gene regulatory networks and cell fate decisions. These simplified models that nonetheless capture the important behaviour of the cell can be useful in understanding fate decision mechanisms and controlling fate decisions artificially e. g. by small molecules. An example of a simple nonlinear rate equation model of the switch controlling HSC differentiation to myeloid or erythroid progenitors has been reported by Andrecut *et al.* recently [8].

Architecture of the self-activation and mutual repression two-gene circuit

HSC differentiate down the erythroid or myeloid pathways depending on the interplay between two key transcription factors, GATA-1 and PU.1. These factors antagonize the expression of the other but stimulate their own expression, a common regulatory switching motif, the bistable latch. This model is represented diagrammatically in figure 10.

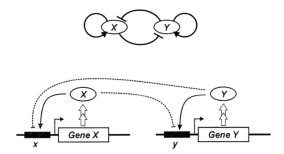

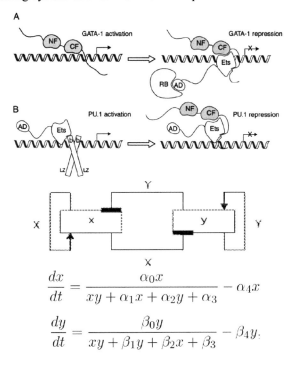

Figure 10. General representation of the bistable switch. Top: coarse-grained circuit scheme for the circuit of two genes X and Y as a dynamical system; bottom: molecular mechanism model amenable for a more detailed chemical reaction kinetics formalism, indicating the variables for the model due to the distinction between genes/promoters (x and y) and the transcription factor proteins (X, Y) [8]. Andrecut *et al.* (2011) *PLoS ONE* **6**(5):e19358.

The mutual antagonism of the two transcription factors, and their autocatalytic stimulation can be represented by a series of rate equations that can be solved numerically. The schematic of this switching system and form of the rate equations is shown in figure 11.

$$\frac{dx}{dt} = \frac{\alpha_0 x}{xy + \alpha_1 x + \alpha_2 y + \alpha_3} - \alpha_4 x$$

$$\frac{dy}{dt} = \frac{\beta_0 y}{xy + \beta_1 y + \beta_2 x + \beta_3} - \beta_4 y.$$

Figure 11. General rate equations describing the mutual antagonism and self-stimulation by the two transcription factors. Andrecut *et al.* (2011) *PLoS ONE* **6**:e19358. *doi:10.1371/journal.pone.0019358;* Winkler *et al.* (2009) *Artif. Life,* **15**(4):411.

When these equations are solved with added noise, this stochastic simulation of the system generates three noisy attractor configurations illustrated in figure 12. These three attractor states can be equated to the uncommitted multipotent attractor, the myeloid attractor with GATA-1 low and PU.1 high, and the erythroid attractor with GATA-1 high and PU.1 low. Interestingly addition of noise to the system generates a manifold of multipotency linking the two committed attractors to the multipotent stem cell state [8]. This may provide a theoretical explanation for the experimentally observed large statistical fluctuations observed in some transcription factors that surprisingly do not trigger a commitment to an associated differentiation pathway.

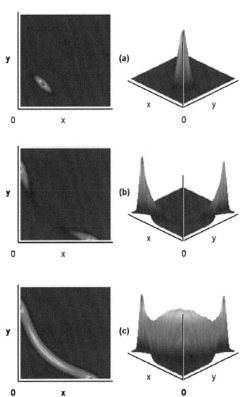

Figure 12. The results of the stochastic simulation of the system for three parameter configurations. **(a)** b.c, **(b)** c.b and **(c)** c = b. (see text for details). Colors (or elevation, respectively) represent the steady state probability distribution (cold-to-warm colors for low-to-high probability for finding the circuit at a given position in the xy-phase plane). Andrecut *et al.* (2011) *PLoS ONE* 6(5):e19358.

Modelling cell fate decisions is one important issue. However, detecting experimentally at an early stage in commitment which decision has been made is also a key and very difficult problem. For example, stem cells must undergo symmetrical division (to become two stem cells) and asymmetric division (to become one stem cell and one progenitor cell) to maintain the stem cell compartment and provide the progenitor cells that generate the required fully differentiated somatic cells (Fig. 13). It is difficult to detect the symmetry of stem cell division. Coupling of gene expression microarray experiments with modern sparse mathematical feature selection methods can help identify markers of the symmetry of cell division.

Molecular Control of (Stem) Cell Fate

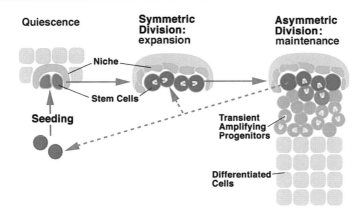

Figure 13. Representation of symmetric and asymmetric stem cell division. *Genes & Dev.* 2007. **21**:3044 – 3060.

Sherley *et al.* recently reported the successful application of this technique to the identification of candidate markers for cell division symmetry [9, 10]. Sparse Bayesian feature selection methods identified a small number of genes from the large number differentially expressed on microarrays derived from the experiments in which the symmetry of cell division was switched artificially by temperature, Zn levels, or p53 expression. Two of the genes identified have striking phenotypes indicative of asymmetric self-renewal in an engineered model cell line. One protein is only down regulated in one sister cell of asymmetric self-renewal divisions as figure 14 illustrates.

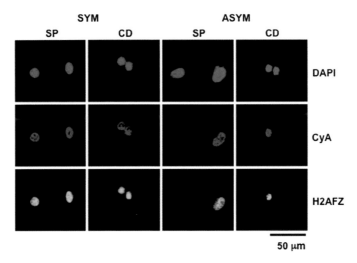

Figure 14. Fluorescently labelled antibodies to one of the cell symmetry marker illustrating localization to one nucleus of a dividing cell (asymmetric division) or both nuclei (symmetric division). Sherley, Smith, Burden, Winkler, *Science,* in preparation.

SMALL MOLECULE MODULATORS OF STEM CELL FATE

Gene regulatory networks that control cell fates decisions are also influenced by external factors such as signaling molecules, cell-cell and cell-matrix interactions, surface compliance/modulus or roughness etc. While native growth factors provide key signals to stem cells via surface expressed receptors that can be used to control stem cell fate, it is not practical to immobilize all of these in engineered products like smart surfaces, implants, or in bioreactors. Additionally, viral transfection methods of reprogramming cells have disadvantages in efficiency and potential risks as a result of genomic integration; limitations that small molecules could circumvent. Small organic molecules that can reliably switch or reprogram cells would have a number of important advantages:

- Chemical and thermal stability

- Control over structure

- Low cost

- Control over tethering position

- Ability to control presentation on surfaces

- Spin-off IP value as drugs or reagents

- Molecular specificity, therefore control over off-target effects

SMALL MOLECULE DRIVERS OF FATE TRANSITIONS

Stem cell fate (survival, proliferation, differentiation, apoptosis) is controlled by a balance between intrinsic internal state of the regulatory network and the presence of external or extrinsic chemical signals provided by e.g.:

- Cytokines

- Growth factors

- Other soluble factors etc.

Cell-matrix and cell-cell interactions via mechanical forces and adhesion factors are also important modulators of fate e.g.:

- Integrins (cell-matrix adhesion)

- Cadherins (cell-cell adhesion)

- Elastic modulus

- Surface patterning

- 2D or 3D environment

- Chemotactic gradients etc.

Surprisingly, given the importance of being able to control stem cell fate decisions, there has been *relatively little* medicinal chemistry research to discover small molecules that can influence cell fate. Notable research efforts include:

- Screening of large chemical libraries for compounds that affect stem cell fate (Figure 15) [11, 12];

- Use of 2i/3i (small molecule antagonists) to maintain 'ground state' pluripotency in mouse [13];

- Rational design of small molecule that mimic protein-protein interactions e.g. small molecule mimetics of cytokines and growth factors that drive adult stem cells down specific differentiation pathways (14 – 16)

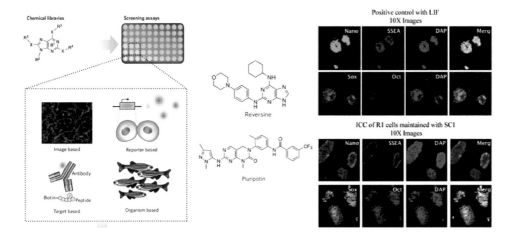

Figure 15. LIF vs. pluripotin for mESC culture. Ding library screening process and two small molecule stem cell effectors discovered. Yue Xu, Yan Shi & Sheng Ding, A chemical approach to stem-cell biology and regenerative medicine

Pluripotent to pluripotent – pluripotency maintenance or self-renewal

Recently, small molecule ligands of the aryl hydrocarbon receptor have been shown to promote self-renewal in HSCs. Regenin1 or SR1 (Fig. 16), and other compounds were identified by screening for compounds that stimulate expansion of CD 34+ cells. Subsequently, it was shown that the small molecule is an antagonist of the aryl hydrocarbon receptor (a nuclear receptor); they did not specifically search for an antagonist. lty and SR1 was shown to stimulate up to 50-fold expansion of CD 34+ cells that maintain full multi-lineage potential and engraft efficiently in mouse transplant models [17].

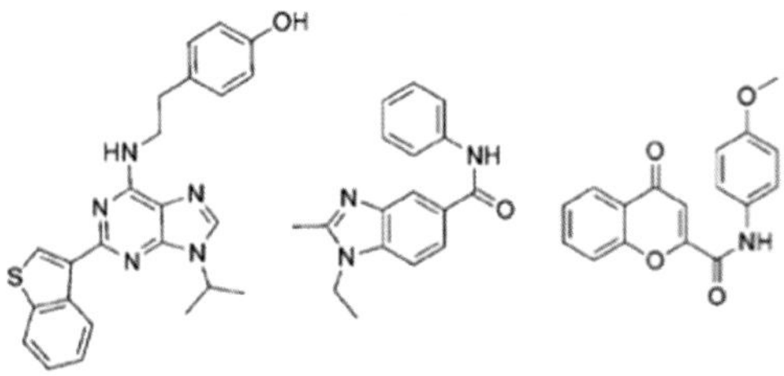

Figure 16. Small molecule antagonists of the AHR.

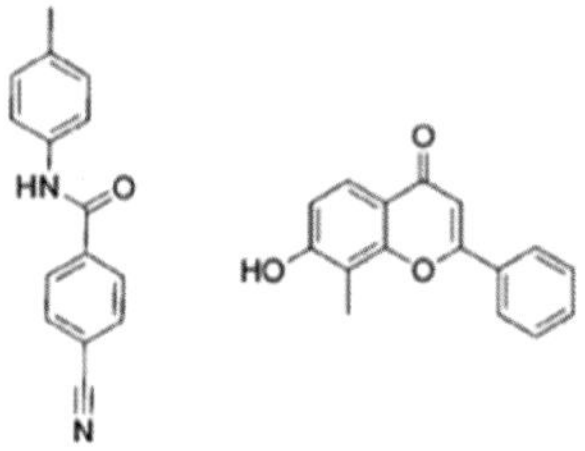

The Smith group in Cambridge reported a combination of two or three small molecule inhibitors (2i/3i) that maintain mouse ES cells in a pristine 'ground' pluripotent state [13]. These three inhibitors are illustrated in figure 17.

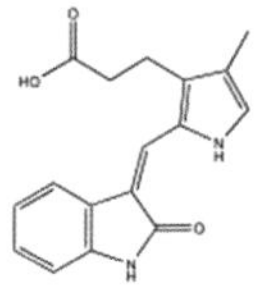

SU5402	CHIR99021	PD184352
fibroblast growth factor receptor antagonist	glycogen synthase kinase 3β inhibitor	mitogen-activated protein kinase kinase 1 inhibitor

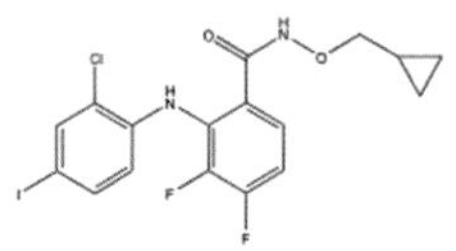

Figure 17. Structure and targets of three small molecule inhibitors that maintain the mES ground state. Ying *et al.* (2008) *Nature* **453**:519; Austin Smith *(WO/2007/ 113505)* Culture medium containing kinase inhibitors and uses thereof.

Pluripotent to somatic – directed differentiation

Mimetics of cytokines, growth factors, or cell adhesion interactions can be used to direct the differentiation of multipotent cells towards desired terminally differentiated cells. Designing small molecules to mimic or block protein-protein interactions is very difficult, but an increasing number of successes are being reported. These small molecules can affect stem cell differentiation directly or when incorporated into artificial bioreactors, but also have intrinsic value as new drugs. Our interest has been in designing mimetic of haematopoietic growth factors and adhesion molecules as robust components of smart surfaces, bioreactors, or as haematological drugs.

The important components of bone marrow niche include [18 – 20]:

- The extracellular matrix: network of osteoblasts, collagen, integrins, fibronectin, aggrecan and link

- growth factors: thrombopoietin, stem cell factor, interleukins (IL-3, IL-6), FLT3L, Notch ligands

We initially focused on designing mimetics of the growth factor thrombopoietin (TPO) that would function as agonists or antagonists of its receptor c-Mpl. We chose TPO because it is a key niche growth factor, there are > 15 known chemical classes of TPO mimicking compounds, although there is no X-ray structure of c-Mpl to aid rational design of ligands.

Thrombopoietin plays a critical role in differentiation of HSCs down the megakaryocytic pathway (Fig. 18) [15].

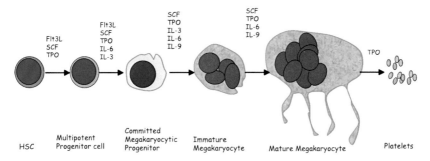

Figure 18. Megakaryocyte pathway promoted by TPO or agonists.

Unusually c-Mpl provides at least two different binding sites for small molecule mimetics. There is a mutation in human and chimpanzee c-Mpl transmembrane domain that places a histidine in the transmembrane helix. This interacts with a large set of small molecules drugs exemplified by Eltrombopag that penetrate the cell membrane, interact with His499 and cause the receptor to dimerise, switch, and facilitate downstream signalling (Fig. 19) [15, 16].

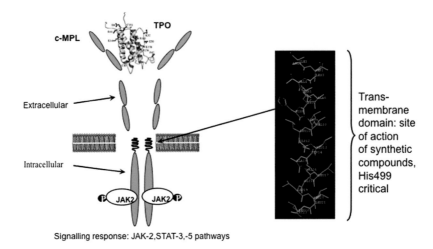

Signalling response: JAK-2,STAT-3,-5 pathways

Figure 19. C-Mpl transmembrane domain showing the location of unique His499.

A selection of the chemical classes that interact with this transmembrane domain in c-Mpl is summarized in figure 20. Molecular modelling methods can locate the low energy conformations (shapes) and common structural alignments that allow the molecular features modulating agonist activity to be elucidated (Fig. 21) [16].

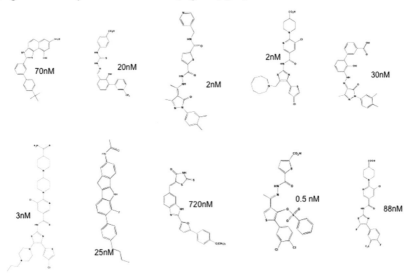

Figure 20. Typical chemical classes of potent small molecule agonists of c-Mpl acting at the transmembrane domain.

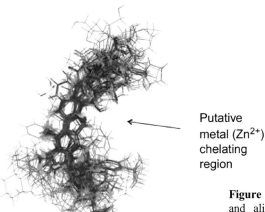

Putative
metal (Zn^{2+})
chelating
region

Figure 21. Common structural motifs and alignments used for modelling TPO agonists [16].

The other potential binding site for small molecule mimetics of TPO is at the extracellular domain of the receptor where the natural ligand TPO binds (Fig. 17). Phage display experiments reported by Cwirla *et al.* identified small peptides that act as agonists and antagonists of c-Mpl [21]. We identified a small conserved epitope (RQW) in these peptides and investigated the role of truncations, mutations, cyclization, dimerization and different dimer linker lengths on agonist activity. This culminated in the discovery of several potent c-Mpl agonists with nanomolar activity in the factor dependent primary cell screen (Fig. 22), in primary CD34+ HSCs, and *in vivo* (Fig. 23) [22]. We also generated the first nanomolar antagonist of c-Mpl.

Peptide	Sequence	%TPO	EC_{50} nM
Dimer	IEGPTLRQWLAARA-K-ARAALWQRLTPGEI | β-Ala	121±15	3.7±2.8
Dimer with pendant K	FmocIEGPTLRQWLAARA-K-ARAALWQRLTPGEIFmoc | β-Ala	105 143	1.0 0.8
Dimer with pendant K and LC-biotin	FmocIEGPTLRQWLAARA-K-ARAALWQRLTPGEIFmoc | β-Ala | K | LC-biotin	102	11
Dimer with pendant K and PEG(2000)-biotin	FmocIEGPTLRQWLAARA-K-ARAALWQRLTPGEIFmoc | β-Ala | K | PEG(2000)-biotin	108 116 113	1.8 3.4 7.5
Truncated Dimer with pendant K	Fmoc-LRQWLAARA-K-ARAALWQRL-Fmoc | β-Ala | K	126	840
Linear-Linear Dimer	AcIEGPTLRQWLAARA**GKG**ARAALWQRLTPGEIAm	85	122
Retroinverso Linear-Linear Dimer	AcIEGPTLRQWLAARA**GKG**IEGPTLRQWLAARAAm	111	3.8
Linear-Truncated Linear	AcIEGPTLRQWLAARA**GKG**ARAALWQRLAm	85	790

Figure 22. Agonist activity of TPO-mimetic peptides.

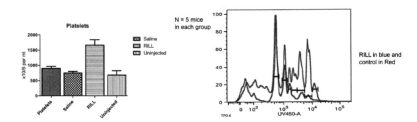

Figure 23. *In vivo* agonist activity of small molecule TPO mimetic. There is a dramatic increase in platelets (left) and megakaryocyte ploidy (right).

Hao *et al.* recently reported small molecules that promote differentiation of ES or iPS cells into cardiomyocytes suitable for cell therapy, cardiac diagnosis, or as screens for new cardiac drug discovery (Fig. 24) [23].

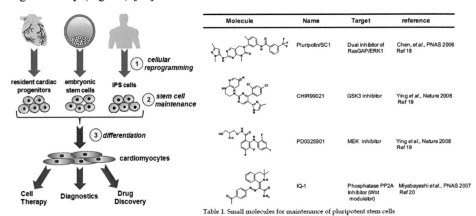

Table 1. Small molecules for maintenance of pluripotent stem cells

Figure 24. Small molecules for SC-based cardiology. Hao *et al.* (2011). Chemical Biology of Pluripotent Stem Cells: Focus on Cardio-myogenesis, in *Embryonic Stem Cells – Recent*

Somatic to pluripotent: Induced pluripotent stem (iPS) cells

Yamanaka was the first to reprogram somatic cells to pluripotency [23, 25]. The main approaches used or proposed to date are (Figure 23):

- Use of retroviruses to transduce mouse fibroblasts with Oct-3/4, SOX2, c-Myc, and Klf4a, the four key pluripotency genes essential for the production of pluripotent stem cells. As c-Myc is oncogenic, 20% of the chimeric mice developed cancer.

- Retroviral mediated reactivation of the same four endogenous pluripotent factors, but selected Nanog as a pluripotency marker. Yamanaka created iPS cells without c-Myc, less efficient, but reduced the risk of cancer.

- Using small compounds that can mimic the effects of transcription factors. While this is not possible yet, the ultimate goal is to discover a cocktail of reprogramming factors and compounds that efficiently and reliably reprogram somatic cells to iPS cells

- Reprogramming through the use of drug-like chemicals activating specific molecular targets, not mimicking transcription factors

- Use of naked DNA, RNA, siRNA and related approaches.

Small molecule compounds may be able to compensate for a reprogramming factor that does not effectively target the genome or fails at reprogramming for another reason, raising reprogramming efficiency. They also avoid the problem of genomic integration, which in some cases contributes to tumorogenesis.

Huangfu *et al.* [26] found that the histone deacetylase inhibitor valproic acid increased reprogramming efficiency 100-fold (compared to Yamanaka's traditional transcription factor method) and proposed that this compound was mimicking the signalling caused by the transcription factor c-Myc without being oncogenic. A similar type of compensation mechanism was proposed to mimic the effects of Sox2. Likewise, Ding *et al.* [27] inhibited histone methyl transferase with the small molecule BIX-01294 in combination with the activation of calcium channels in the plasma membrane to increase reprogramming efficiency. It is foreseeable that such experiments will continue to find small compounds that improve efficiency rates.

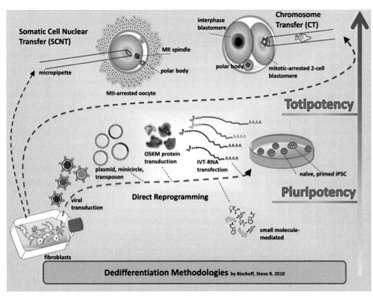

Figure 25. Gene, viral and small molecule methods for dedifferentiation and somatic cell reprogramming to pluripotency. From: *Advances in Pluripotent Stem Cell-Based Regenerative Medicine*, Atwood (Ed.), ISBN: 978-953-307-198-5, (23)

Somatic to somatic

The recent recognition of the plasticity of cell identity and the nascent capabilities to reprogram some cells into other types leads inevitably to speculation on whether any cell can in principle be reprogrammed into anything else. This is termed direct reprogramming and is attracting considerable interest. Li *et al.* [28] have recently demonstrated that transient overexpression of reprogramming factors in fibroblasts leads to the rapid generation of epigenetically activated cells (unstable intermediate populations). These can be coaxed to back into various differentiated state(s), ultimately giving rise to fully differentiated cells entirely distinct from the starting population, as figure 26 illustrates.

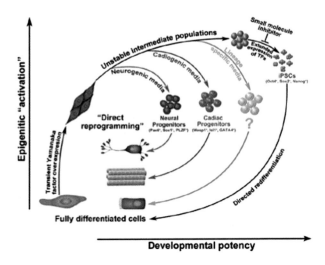

Figure 26. Direct reprogramming of somatic cells to other cell phenotypes using small molecules. Li *et al.* (2012) *Stem Cells* **30**:61 – 68.

A related report by Kim *et al.* [29] outlined a strategy for discovery of small molecules with potential for limb regeneration. They also involved a high throughput screen of chemical libraries to identify molecules capable of altering proliferation and gene expression profiles in urodele amphibian skeletal muscle cells. The small molecules BIO (glycogen synthase-3 kinase inhibitor), lysophosphatidic acid (pleiotropic activator of G-protein-coupled receptors), SB203580 (p38 MAP kinase inhibitor), or SQ22536 (adenylyl cyclase inhibitor) induced proliferation. These proliferating cells were multipotent and possessed a markedly different gene expression pattern than lineage-restricted myoblasts. Genes related to signal transduction and differentiation were particularly affected (Fig. 27). Some molecules were found to promote skeletal muscle dedifferentiation and differentiation into alternate cell types.

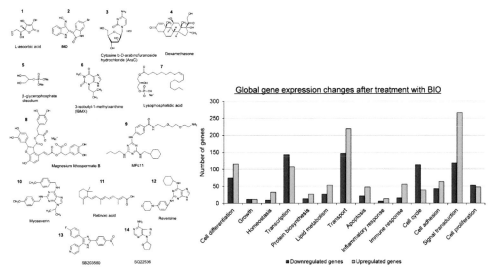

Figure 27. Small molecules discovered by chemical library screens that promote dedifferentiation of muscle cells into multipotent cells with markedly different gene expression profiles. Kim *et al.* (2012) *ACS Chem. Biol.* ASAP.

Aberrant pluripotent to somatic or death

The recent recognition that aberrant stem cell, cancer stem cells, play an important role on many forms of cancer has provided a promising new approach to therapy. Several groups are investigating targeting cancer stem cells directly using small molecules [30]. While this research is at a very early stage, if cancer stem cells can be selectively identified, killed, or forced to differentiate treatments may be more effective and recurrence of drug resistant tumour less likely. Ischenko *et al.* [30] reported that long-term treatment of gliomasphere cells with the cyclopamine (Fig. 26) alone killed all cancer stem cells in culture, and induced the regression of glioma tumors established from the gliomasphere cells in nude mice *in vivo*, without detectable secondary effects.

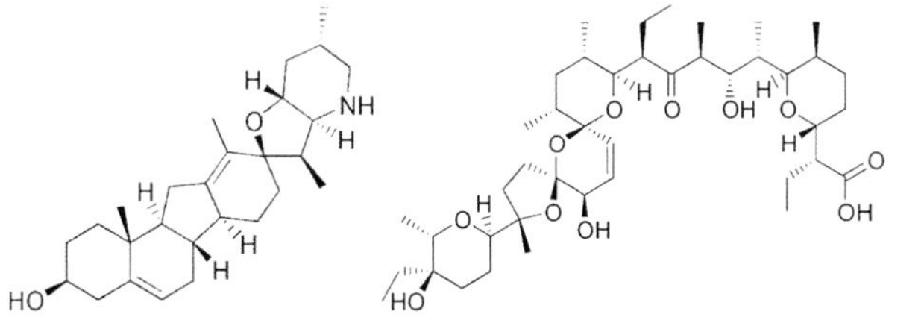

Figure 28. Cyclopamine (left) and Salinomycin (right). Ischenko *et al.* (2008) *Current Medicinal Chemistry*/Wiki Commons.

Gupta *et al.* [31] also adopted a small molecule library screening approach to discover compounds targeting breast cancer stem cells. They reported that salinomycin reduces the proportion of breast cancer stem cells in culture by > 100-fold compared to another cytotoxic cancer drug paclitaxel, Salinomycin also inhibits mammary tumour growth *in vivo* and reduce the expression of breast cancer stem cell-related genes.

PERSPECTIVE

Clearly, it is early days in the discovery, design, and use of small organic molecules and peptides to control cell fate. Research over the past five years has been very encouraging. It is likely that reprogramming of somatic cells to pluripotency will be achievable largely or entirely by small molecules in short to medium time frames. There is increasing evidence that physical and chemical cues can control cell fate, and that transdifferentiation of one somatic cell type to another without isolating any multipotent intermediate may also be possible in the medium term. There are good prospects that this will be achieved using a combination of topographical, mechanical and/or chemical cues. This has interesting implications for tissue replacement and engineering, providing hope that many injuries, illnesses and congenital defects can be treated effectively where current therapies do not exist or provide a poor outcome. One of the most exciting prospects is selective targeting of cancer stem cells by small molecule drugs. If this is realized, treatments for cancers could improve dramatically because the underlying cause of many tumours is eliminated. Time will tell how many of these exciting possibilities come to fruition. Finally, while this chapter was being finalized, it was announced that the Nobel Prize in Medicine had been awarded to Gurdon and Yamanaka in recognition of their discovery of cellular reprogramming.

ACKNOWLEDGEMENTS

The author gratefully acknowledges the very substantial contributions of the following research staff at CSIRO: Laurence Meagher, David Haylock, Susie Nilsson, Andrew, Riches, Jacinta White, Anna Tarasova, Cheang By, Jessica Andrade, Teresa Cablewski, Glenn Condie, Jerome Werkmeister. The author is also grateful to Dr. David Haylock for his careful reading of the text and helpful suggestions.

REFERENCES

[1]   Roeder, I., Loeffler, M. (2002) A novel dynamic model of hematopoietic stem cell organization based on the concept of within-tissue plasticity. *Experimental Hematology* **30**(8):853 – 61.
doi: 10.1016/S0301-472X(02)00832-9.

[2]   Takahashi, K., Yamanaka, S. (2006) Induction of pluripotent stem cells from mouse embryonic and adult fibroblast cultures by defined factors. *Cell* **126**(4):663 – 76.
doi: 10.1016/j.cell.2006.07.024.

[3]   Roeder, I., Radtke, F. (2009) Stem cell biology meets systems biology. *Development* **136**(21):3525 – 30.
doi: 10.1242/dev.040758.

[4]   Huang, S., Eichler, G., Bar-Yam, Y., Ingber, D.E. (2005) Cell fates as high-dimensional attractor states of a complex gene regulatory network. *Physical Review Letters* **94**(12).
doi: 10.1103/PhysRevLett.94.128701.

[5]   Kauffman, S.A. (1969) Metabolic Stability and Epigenesis in Randomly Constructed Genetic Nets. *Journal of Theoretical Biology* **22**(3):437.
doi: 10.1016/0022-5193(69)90015-0.

[6]   Waddington C. (1940) The strategy of the genes. Cambridge. Cambridge University Press.

[7]   Macarthur, B.D., Ma'ayan, A., Lemischka, I.R. (2009) Systems biology of stem cell fate and cellular reprogramming. *Nature Reviews Molecular Cell Biology* **10**(10):672 – 81.
doi: 10.1038/nrm2766.

[8]   Andrecut, M., Halley, J.D., Winkler, D.A., Huang, S. (2011) A General Model for Binary Cell Fate Decision Gene Circuits with Degeneracy: Indeterminacy and Switch Behavior in the Absence of Cooperativity. *PLoS One* **6**(5).
doi: 10.1371/journal.pone.0019358.

[9]   Huh, Y.H., Noh, M., Burden, F.R., Chen, J.C., Winkler, D.A., Sherley, J.L. (2012) Sparse Feature Identification of H2A.Z as a Biomarker for Distributed Stem Cells. *Science*, submitted.

[10] Noh, M., Smith, J.L., Huh, Y.H., Sherley, J.L.. (2011) A Resource for Discovering Specific and Universal Biomarkers for Distributed Stem Cells. *PLoS One* **6**(7).
doi: 10.1371/journal.pone.0022077.

[11] Xu, Y., Shi, Y., Ding, S. (2008) A chemical approach to stem-cell biology and regenerative medicine. *Nature* **453**(7193):338–44. doi: 10.1038/nature07042.

[12] Xiong, W., Gao, Y., Cheng, X., Martin, C., Wu, D., Yao, S. *et al.* (2009) The use of SC1 (Pluripotin) to support mESC self-renewal in the absence of LIF. *J. Vis. Exp.* (33).

[13] Ying, Q.L., Wray, J., Nichols, J., Batlle-Morera, L., Doble, B., Woodgett, J. *et al.* (2008) The ground state of embryonic stem cell self-renewal. *Nature* **453**(7194):519. doi: 10.1038/nature06968.

[14] Margulies, D., Opatowsky, Y., Fletcher, S., Saraogi, I., Tsou, L.K., Saha, S. *et al.* (2009) Surface Binding Inhibitors of the SCF-KIT Protein-Protein Interaction. *Chembiochem* **10**(12):1955–8. doi: 10.1002/cbic.200900079.

[15] Tarasova, A., Haylock, D., Winkler, D.A. (2011) Principal signalling complexes in haematopoiesis: Structural aspects and mimetic discovery. *Cytokine & Growth Factor Reviews* **22**(4):231–53. doi: 10.1016/j.cytogfr.2011.09.001.

[16] Tarasova, A., Winkler, D.A. (2009) Modelling Atypical Small-Molecule Mimics of an Important Stem Cell Cytokine, Thrombopoietin. *Chemmedchem* **4**(12):2002–11. doi: 10.1002/cmdc.200900340.

[17] Boitano, A.E. (2011) Aryl hydrocarbon receptor antagonists promote the expansion of human hematopoietic stem cells (September, pg 1345, 2010). *Science* **332**(6030):664.

[18] Ellis, S.L., Nilsson, S.K. (2012) The location and cellular composition of the hemopoietic stem cell niche. *Cytotherapy* **14**(2):135–43. doi: 10.3109/14653249.2011.630729.

[19] Nilsson, S.K. (2010) Understanding the Hsc Niche. *Experimental Hematology* **38**(9):S108.

[20] Haylock, D.N., Nilsson, S.K. (2005) Stem cell regulation by the hematopoietic stem cell niche. *Cell Cycle* **4**(10):1353–5. doi: 10.4161/cc.4.10.2056.

[21] Cwirla, S.E., Balasubramanian, P., Duffin, D.J., Wagstrom, C.R., Gates, C.M., Singer, S.C. *et al.* (1997) Peptide agonist of the thrombopoietin receptor as potent as the natural cytokine. *Science* **276**(5319):1696–9. doi: 10.1126/science.276.5319.1696.

[22]   Tarasova, A., Meagher, L., Haylock, D., White, J., Winkler, D.A. (2012) Small potent peptide agonists of thrombopoiesis. *J. Med. Chem.*, in press.

[23]   Hao, J., Sawyer, D.B., Hatzopoulos, A.K., Hong, C.C. (2011) Recent Progress on Chemical Biology of Pluripotent Stem Cell Self-renewal, Reprogramming and Cardiomyogenesis. *Rec. Pat. Regen. Med.* **1**(3):263 – 74.

[24]   Yamanaka, S. (2007) Strategies and new developments in the generation of patient-specific pluripotent stem cells. *Cell Stem Cell* **1**(1):39 – 49.
doi: 10.1016/j.stem.2007.05.012.

[25]   Yamanaka, S. (2008) Pluripotency and nuclear reprogramming. *Philosophical Transactions of the Royal Society B-Biological Sciences* **363**(1500):2079 – 87.
doi: 10.1098/rstb.2008.2261.

[26]   Huangfu, D.W., Maehr, R., Guo, W.J., Eijkelenboom, A., Snitow, M., Chen, A.E. *et al.* (2008) Induction of pluripotent stem cells by defined factors is greatly improved by small-molecule compounds. *Nature Biotechnology* **26**(7):795 – 7.
doi: 10.1038/nbt1418.

[27]   Ding, S. (2008) A chemical approach to stem cell biology. *Cell Research* **18.**

[28]   Li, W.L., Jiang, K., Ding, S. (2012) Concise Review: A Chemical Approach to Control Cell Fate and Function. *Stem Cells* **30**(1):61 – 8.
doi: 10.1002/stem.768.

[29]   Kim, W.H., Jung, D.W., Kim, J., Im, S.H., Hwang, S.Y., Williams, D.R. (2012) Small Molecules That Recapitulate the Early Steps of Urodele Amphibian Limb Regeneration and Confer Multipotency. *ACS Chem. Biol.* **7**(4):732 – 43.
doi: 10.1021/cb200532v.

[30]   Ischenko, I., Seeliger, H., Schaffer, M., Jauch, K.W., Bruns, C.J. (2008) Cancer Stem Cells: How can we Target them? *Curr. Med. Chem.* **15**(30):3171 – 84.
doi: 10.2174/092986708786848541.

[31]   Gupta, P.B., Onder, T.T., Jiang, G.Z., Tao, K., Kuperwasser, C., Weinberg, R.A. *et al.* (2009) Identification of Selective Inhibitors of Cancer Stem Cells by High-Throughput Screening. *Cell* **138**(4):645 – 59.
doi: 10.1016/j.cell.2009.06.034.

Beilstein Bozen Symposium on Molecular Engineering and Control
May 14th – 18th, 2012, Prien (Chiemsee), Germany

PLP-dependent Enzymes: a Powerful Tool for Metabolic Synthesis of Non-canonical Amino Acids[1]

Martino L. di Salvo[1]*, Nediljko Budisa[2], and Roberto Contestabile[1]

[1]Dipartimento di Scienze Biochimiche "A. Rossi Fanelli",
Sapienza Università di Roma,
Via degli Apuli 9, 00185 Roma, Italy.

[2]Institut für Biologische Chemie, Biocatalysis Group, Technische Universität Berlin,
Müller-Breslau-Str. 10, 10623 Berlin, Germany.

E-Mail: *martino.disalvo@uniroma1.it

Received: 5th June 2013 / Published: 13th December 2013

An Overview on PLP-dependent Enzymes

Pyridoxal 5'-phosphate (PLP), the biologically active of vitamin B_6, was first identified in the mid-forties as the cofactor for the transamination reaction. Since then, PLP-dependent enzymes have been the focus of extensive biochemical research. The interest aroused by these enzymes is due to their unrivalled catalytic versatility and their widespread involvement in cellular metabolism. As a matter of fact, PLP acts as cofactor in more than 160 different enzymes classified by the Enzyme Commission, representing 4% of all known cellular catalytic activities [1]. PLP-dependent enzymes serve vital roles in all living organisms and catalyze a number of diverse chemical reactions, such as transamination, decarboxylation, racemization, carbon-carbon bond cleavage and formation. PLP-dependent activities are involved in essential biosynthetic pathways including glucose and lipid metabolism,

---

[1] Non-canonical amino acids (ncAA): amino acids that do not participate to protein translation, i. e. are not genetically encoded. We will use the term ncAA for both naturally and synthetically generated amino acids that are used to expand the scope of protein synthesis

amino acid metabolism, heme and nucleotide synthesis, and neurotransmitter production [2, 3]. As a consequence of their crucial metabolic relevance, a number of these enzymes are widely recognized drug targets [4].

The mechanism of many B_6 enzymes has been studied extensively over the last 50 years, with respect to structure, function, substrate and reaction specificity. Because of their catalytic versatility, in the recent years PLP-dependent enzymes have acquired extraordinary importance in biotechnology, to be exploited in the semi-synthetic production of compounds for medical and industrial use.

PLP chemistry

PLP resembles benzaldehyde in its structure. However, the properties of the carbonyl group are modified by the presence both of an adjacent hydroxyl group and ring nitrogen in the *para* position (Fig. 1), whose protonation state has profound effects on the cofactor's reactivity. These features of PLP were recognized to be of fundamental importance for catalysis in the early model studies carried out on the cofactor alone and on its analogues [5].

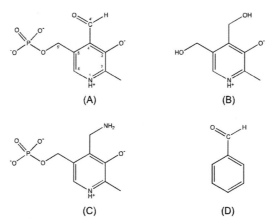

(A) (B)

(C) (D)

Figure 1. Structure of pyridoxal 5'-phosphate (PLP) and related compounds. **(A)** pyridoxal 5'-phosphate, the active form of vitamin B6 used as enzyme cofactor; **(B)** pyridoxine (PN), the vitamin B6 form that is most commonly given as dietary supplement; **(C)** pyridoxamine 5'-phosphate (PMP), the other natural occurring vitamin B6-derived catalyst; **(D)** benzaldehyde.

Two basic, interdependent chemical properties of PLP are involved in catalysis. First, its ability to form imines with primary amino groups through its aldehyde group, and second its facility to stabilize carbanionic intermediates that develop by heterolytic cleavage of chemical bonds. Typically, the cofactor forms two types of imine. In the free enzyme an *internal aldimine* is formed with a lysine side chain. *External aldimines* are formed with amino acids and closely related compounds (Fig. 2). Both types of imine react reversibly with primary amines in a transaldimination reaction (step 2 to 5 in Fig. 2), with formation of a *geminal diamine* intermediates, allowing binding of substrates and release of products (forward and reverse of step 3 in Fig. 2).

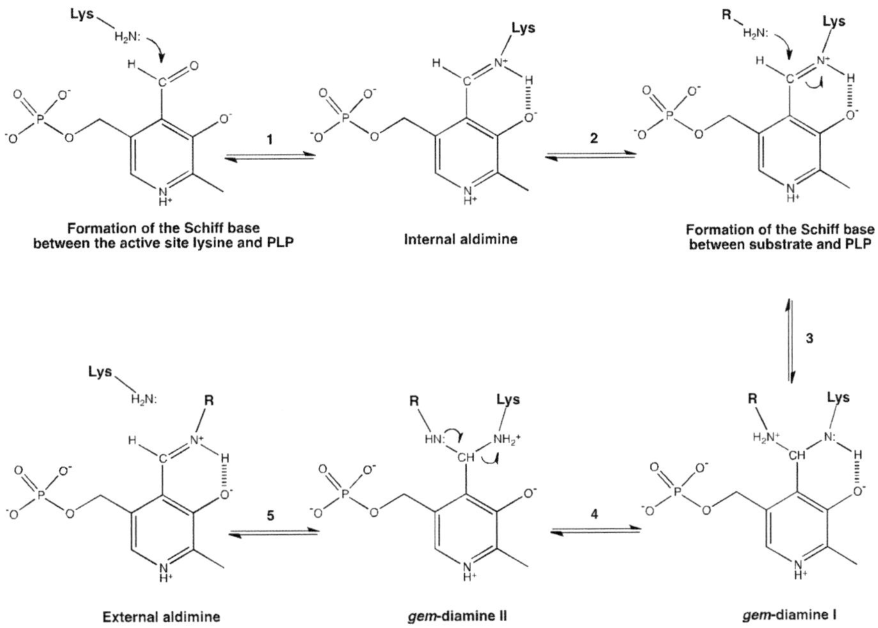

Figure 2. PLP reacts reversibly with primary amines to form imines; **(1)** formation of the *internal aldimine* with the active site lysine residue; **(2 to 5)** transaldimination reaction: formation of the *external aldimine* with the substrate amino acid, with formation of *geminal diamine* intermediates. The transaldimination reaction between the internal and the external aldimines allows substrate binding and product release.

The reactions described so far are characteristic of any carbonyl compound. PLP is unusual because of the high efficacy with which they are accomplished, that relies on the properties of its heteroaromatic pyridine ring. The electrophilicity of C4' is in fact greatly enhanced through the electron withdrawing effect exerted by the protonated pyridinium nitrogen (Fig. 3A). Moreover, the phenoxide anion at C3 stabilizes the protonated state of the imine nitrogen, by means of resonance and hydrogen bonding (Fig. 3B), that further increases the electrophilic character of C4' and easily accommodates an electron lone pair during trans-aldimination. At the same time, these coulombic (inductive and field) and resonance effects, the former strictly dependent on the latter, are responsible for the second catalytic property of PLP, its action as an "electron sink". The protonated N1 of the cofactor in the external aldimine withdraws electrons from Cα of substrates. This electron deficit, further increased if the imine nitrogen is itself protonated, produces a polarization, and therefore a weakening, of σ bonds to Cα.

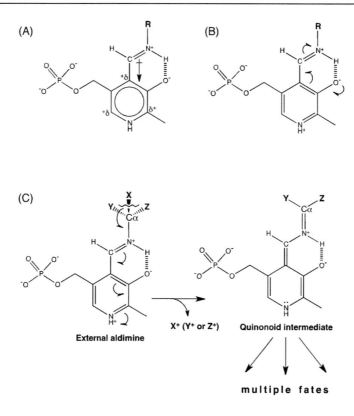

Figure 3. (A) Coulombic and **(B)** resonance effects involved in PLP-dependent catalysis. **(C)** Formation of the *quinonoid intermediate* upon the heterolytic cleavage of one of the bonds around Cα. PLP acts as an electron sink in the formation of the quinonoid intermediate, the common steps in all PLP-catalyzed reactions.

The extensive conjugation of the π-electrons of the ring, which extends to the imine bond and to the oxy substituent at C3, very efficiently delocalizes the net negative charge arising from the heterolytic cleavage of these bonds. The protonated ring nitrogen plays a major role in stabilizing the net negative charge (Fig. 3C). Besides the gain in delocalization energy, which is a thermodynamic aspect, an additional, kinetic consideration must be taken into account. This focuses on the transition state for bond breaking. An important factor in the activation of σ bonds by a π system is the stereochemical arrangement. Dunathan [6] pointed out that if the gain in delocalization energy is to aid the bond breaking process, the transition state must assume a geometry that approaches that of the coplanar product, i.e. which places the bond to be broken in a plane perpendicular to that of the π system. Figure 4 shows this basic concept and, at the same time, explains how reaction specificity may be controlled.

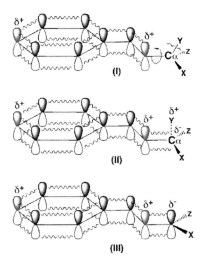

Figure 4. Schematic representation of the π-orbital framework in a PLP-substrate external aldimine. The minimum level of energy of the transition state in the cleavage of a bond to Cα is that in which the π-orbital that is forming is in conjugation with the π-system of the pyridine ring **(II)**, the nitrogen atoms are displayed in blue. This situation takes place when the σ-bond to be cleaved is perpendicular to the pyridine ring, as in **(I)**. This geometry of the transition state, which approaches that of the quinonoid product **(III)**, allows the extension of the electronic delocalisation to the bond that is breaking, stabilising the developing negative charge. As a consequence, the bond to Cα perpendicular to the pyridine ring is the most labile because of its overlap with the π-system. Rotation around the bond between Cα and the imine nitrogen **(I)** determines which of the three bonds to Cα will be cleaved and therefore controls the reaction specificity ('Dunathan's hypothesis' [6]).

The cofactor is able to catalyze multiple reactions in the complete absence of enzymes (although at much slower rates than those characteristic of enzymes) such as transamination of amino acids [5], racemization [7] and, with serine, α,β-elimination [8]. The rates of reactions catalyzed by pyridoxal phosphate alone were found to be accelerated by metal ions, such as Cu^{2+} and Al^{3+} [8]. Since these ions are known to chelate with Schiff's bases of the type formed with PLP, it was concluded that their action is that of providing coplanarity between the imine and the pyrimidine ring. The role played by metal ions in model systems is carried out by the hydrogen bond between O3 and the imine nitrogen in the enzymes. The function of the phosphorylated hydroxymethyl substituent at C5 is clearly to provide a firm anchor to the coenzyme. PLP-dependent enzymes bind the cofactor phosphate group through a similar set of interactions. This common recognition pattern was named "phosphate-binding cup" [9]. The methyl substituent at C2 appears not to have a function and may simply arise from metabolic requirements [10].

It is worth of note that the pool of free PLP *in vivo* must be maintained at a very low level, to prevent toxic build-up. In fact, PLP being a very reactive aldehyde, easily combines with amines and thiols in the cell. This characteristic has been related, for example, to the neurotoxic effect of excess consumption of vitamin B_6 [11]. In eukaryotic cells, the concentration of free PLP is maintained as low as $1\,\mu M$. There must then be a tight regulation control on PLP homeostasis, that would insure the presence of enough cofactor for all PLP-dependent enzymatic activity in the cell. Failure to maintain the correct tuning between PLP biosynthesis, degradation, and delivering to newly formed PLP-enzymes might end up in poor cell growth or vitamin B_6-associated pathogenesis such as severe neurological disorders (epilepsy, schizophrenia, Alzheimer's and Parkinson's diseases) [12].

Mechanisms of reaction

In spite of all the different reaction carried out by PLP-dependent enzymes, it is possible to envisage a series of common mechanistic features characteristic of PLP-catalyzed reactions. In the early 1950 s, a general mechanism of reaction was postulated based on considerations on the various known PLP-dependent enzymes and model studies on the cofactor alone [5]. All of the many different reactions begin with conversion of the PLP-enzyme internal aldimine to the PLP-substrate external aldimine and the consequent heterolytic cleavage of one of the three bonds to Cα (Fig. 3C and 5). This latter step represents the first point of diversification. The three distinct carbanions formed are called quinonoid intermediates and react further in multiple paths that themselves branch in a series of consecutive steps, invariably ending up in either an aldimine or a ketimine adduct which is transaldiminated or hydrolyzed in order to release the related product. It follows that an exposition of the essential steps in the mechanisms of the varied types of reactions can be approached progressively, starting from the above mentioned common quinonoid intermediates and moving towards increasing levels of diversification (Fig. 5).

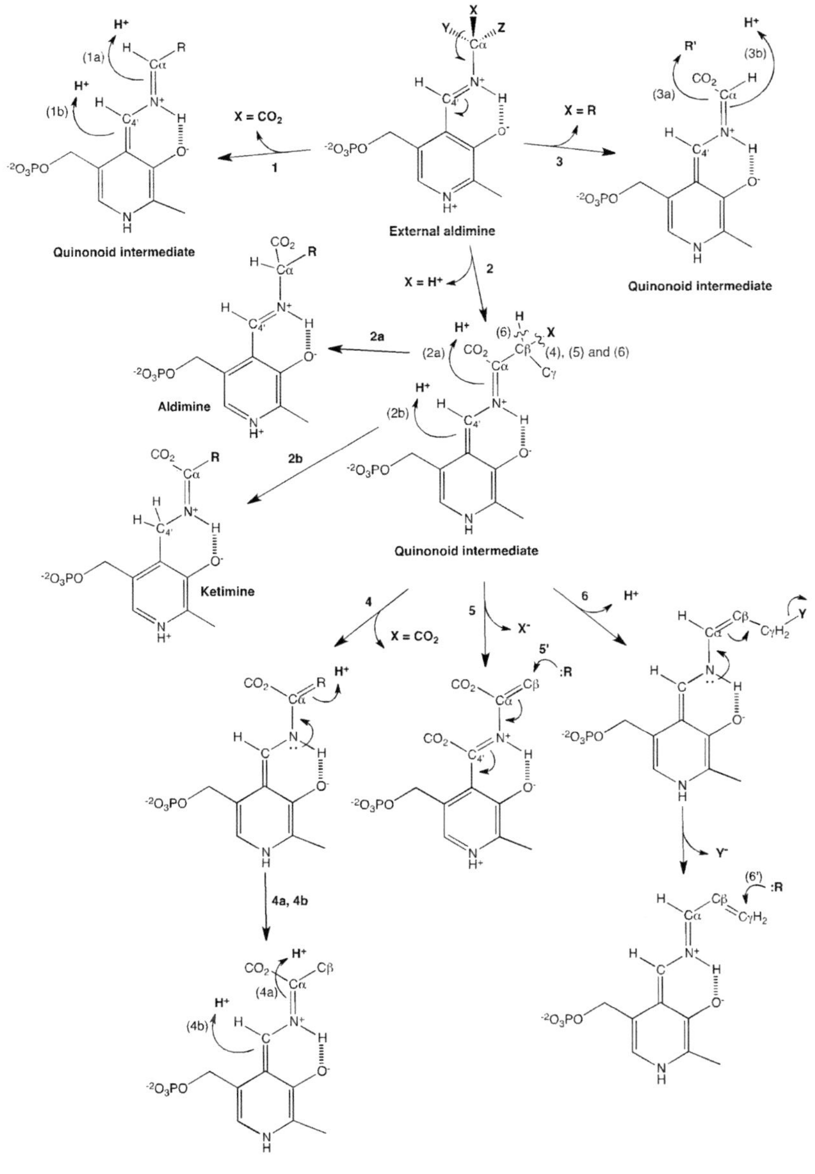

Figure 5. Schematic representation of the various types of reaction catalyzed by PLP-dependent enzymes.

Legend:

(1) Reactions proceeding through elimination of CO_2 from $C\alpha$:
    (1a) α-decarboxylation
    (1b) transaminating decarboxylation

(2) Reactions proceeding through deprotonation of $C\alpha$:
    (2a) aldimine formation: racemization
    (2b) ketimine formation: transamination

(4) β-decarboxylation
    (4a) aldimine formation: β-decarboxylation
    (4b) ketimine formation: transaminating β-decarboxylation
(5) β-elimination
(5') β-synthesis
    (5'a) aldimine formation: production of a new β-substituted amino acid
    (5'b) ketimine formation: production of a new β-substituted keto acid
(6) γ-elimination
(6') γ-synthesis
(3) Reactions proceeding through elimination of the side chain:
    (3a) α-synthesis: production of a β-hydroxyamino acid
    (3b) retroaldolic cleavage: releasing of glycine and an aldehyde

Pyridoxal phosphate is clearly a tool in the "hands" of proteins. The efficacy of each enzyme relies on its ability to accelerate certain reactions selectively at the expense of others, thereby making its way through the intricate maze of the many possible chemical transformations. The outstanding catalytic diversity of PLP-dependent enzymes arises from modulation and enhancement of the coenzyme intrinsic chemical properties by the surrounding polypeptide matrix.

On the basis of the available structural information, it is generally accepted that PLP-dependent enzymes originated very early in the evolution (before the three biological kingdoms diverged, some 1500 million years ago) from different protein ancestors, which generated at least five independent families, each corresponding to a different fold type [13]. These families have been named from their more representative enzyme. The aspartate aminotransferase family corresponds to fold type I and contains the majority of structurally determined PLP-dependent enzymes. The tryptophan synthase β-subunit family corresponds to the fold type II; the bacterial alanine racemase family corresponds to the fold type III; the D-amino acid aminotransferase family corresponds to the fold type IV; the glycogen phosphorylase corresponds to fold type V. The glycogen phosphorylase family seems to be a totally unrelated example, in which PLP is used in a completely different way. In that case, in fact, the reaction chemistry occurs through a general acid-base mechanism that involves the phosphate moiety of the cofactor. In all other PLP-dependent enzyme families, the phosphate group acts only as an anchor to tightly bind the cofactor to the folded polypeptide, whereas the chemistry is carried out by the pyridine ring moiety.

The observation that PLP-dependent enzymes have multiple evolutionary origins is very interesting: the common mechanistic features of PLP-enzymes are not then accidental historical traits but may reflect evolutionary necessities based on PLP chemistry.In support to this hypothesis is the observation of several examples of convergent evolution, provided by enzymes catalyzing the same reactions but belonging to different fold types, concerning both similarities of cofactor binding sites and catalytic mechanisms [14].

Nowadays, more than a hundred crystal structures of PLP-dependent enzymes, most of them solved in the last 5 years, are accessible from databases.

Catalytic versatility and promiscuity

Enzymes are generally believed to be very specific in their action, i. e. to be endowed with a strict reaction and substrate specificity. However, as a matter of fact, many enzymes are able to catalyze more than one reaction, often using different substrates. These enzymes are called generalist in opposition to specialist enzymes, which evolved to catalyze one reaction on a unique primary substrate. They represent a significant portion of the total enzymes in a living cell and play more than one physiologically important role [15]. Besides this natural eclecticism, many enzymes when acting on their physiological substrate are prone to make "mistakes" and catalyze measurable side reactions. Moreover, *in vitro* and when presented with substrates analogues (also unnatural compounds), numerous enzymes show activities that are not part of the organism's physiology. This capability of an enzyme to catalyze alternative reactions (often different types of reactions with different substrates) is referred to as catalytic promiscuity [16, 17]. Catalytic promiscuity is believed to have played a fundamental role in divergent evolution and diversification of catalytic properties. Ancestral enzymes were probably able to catalyze a range of different reactions. Gene duplication and evolutionary pressure may have worked to shape enzymes' active sites so as to confer narrower substrate and reaction specificity.

Catalytic promiscuity offers very important biotechnological opportunities. Many enzymes may be used as such for the enzymatic synthesis of unnatural compounds. Alternatively, they may be engineered so as to catalyze novel reactions. In this context, PLP-dependent enzymes represent a particularly interesting source of promiscuous catalysts. PLP itself is able to catalyze multiple reactions, although at a much slower rate compared to the enzymes that use it as cofactor. This exceptional catalytic versatility is often responsible for a pronounced catalytic promiscuity. For example, the physiological role of aspartate aminotransferase (Asp-AT) is to catalyze the reversible transamination that converts L-aspartate and α-ketoglutarate into oxaloacetate and L-glutamate. Asp-AT catalyzes this reaction almost exclusively with its own natural substrates. L-enantiomers of tyrosine, phenylalanine and alanine undergo transamination $3-5$ orders of magnitude slower than the dicarboxylic substrates. However, the above-mentioned compounds are also racemized by Asp-AT with a rate of $10^{-5} - 10^{-6}$ s^{-1} [18, 19].

Moreover, when presented with appropriate, unnatural substrate analogues, Asp-AT efficiently catalyzes β-elimination reactions [20]. This apparent "imperfection" in the catalytic machinery of the enzyme is actually due to the presence of a good leaving group at Cβ of such analogues. After deprotonation of the substrate-cofactor aldimine (step 2 in Fig. 5 and step 1 in Fig. 6) Cα and C4' of the quinonoid intermediate have a marked nucleophilic character. When acting on physiological substrates Asp-AT stereospecifically reprotonates either Cα or C4' to form the related external aldimines (reverse of step 1 and step 2a in Fig. 6). A nucleophilic attack by Cα on Cβ is instead the prevalent reaction if a good leaving group exists at the latter carbon (step 2b in Fig. 6).

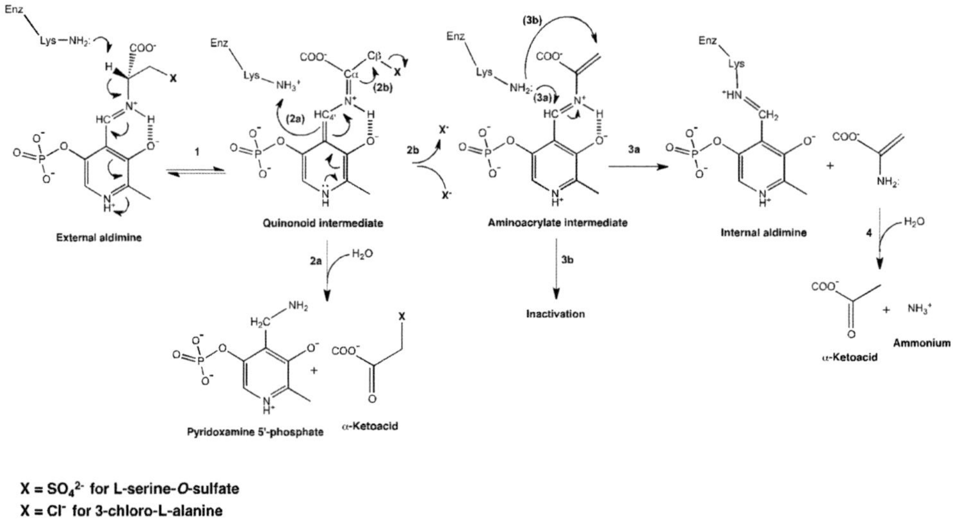

X = SO₄²⁻ for L-serine-*O*-sulfate
X = Cl⁻ for 3-chloro-L-alanine

Figure 6. Mechanism of reaction of Asp-AT-catalyzed β-elimination **(2b)** and transamination **(2a)**. Details of the mechanism are described in the test.

The subsequent transaldimination then releases the aminoacrylate intermediate, which spontaneously hydrolyses to form ammonium and the related keto acid (step 3a and 4 in Fig. 6). Alternatively, the active site lysine reacts with the aminoacrylate intermediate inactivating the enzyme (step 3b in Fig. 6). As an example, L-serine *O*-sulphate [21] and 3-chloro-L-alanine [22] are both rapidly converted to pyruvate, ammonia and SO_4^{2-} or Cl⁻ respectively. The ability of Asp-AT to catalyze β-elimination of α-amino acids that have a good leaving group at Cβ has been exploited in the synthesis of unnatural amino acids such as sulfocysteine by the inclusion of appropriate nucleophiles as co-substrates, which attack the aminoacrylate intermediate forming a stable product [23]. Although aminoacrylate is abnormal to the aminotranferase reaction, it is, together with aminocrotonate (same structure of aminoacrylate with a methyl group at Cβ), an intermediate for many PLP-dependent enzymes that catalyze β-elimination reactions. Among them are serine and threonine dehydratases (elimination of OH⁻, as H_2O), bacterial tryptophanase (elimination of indole) and alliinase from onions and garlic (elimination of 1-propenylsulphenic acid, the lachrymator formed on crushing these herbs). It is also a natural intermediate in synthetic reactions such as that catalyzed by tryptophan synthase [25]. Some of the above mentioned enzymes have been exploited for the synthesis of non-canonical amino acids (ncAA) based on the reaction of the aminoacrilate intermediate with nucleophilic compounds (see below).

Remarkably, some PLP-dependent enzymes catalyze alternative reactions that correspond to the main reaction catalyzed by evolutionary related enzymes. Serine hydroxymethyltransferase (SHMT), an enzyme with exceptionally broad reaction specificity, is an interesting example of this phenomenon and of catalytic promiscuity. On top of the hydroxymethyltransferase reaction, SHMT is able to catalyze decarboxylation, aldol cleavage, transamination and racemization reactions, using different natural or unnatural substrates. Two other PLP-dependent enzymes, threonine aldolase (TA) and fungal alanine racemase (AlaRac), turned out to be very close to SHMT from the point of view of structural similarities and catalytic properties, being able to catalyze the same reactions, although with different efficiencies [25, 26].

NON-CANONICAL AMINO ACIDS FOUND IN NATURAL COMPOUNDS

As pointed out in the previous chapter, due to their extraordinary versatility and catalytic efficiency PLP-dependent enzymes can be envisaged as powerful tools for the semi-synthetic production of non-canonical amino acids and their related compounds. The following sections review some of the most common non-canonical amino acids found in natural compounds (this section), whose production can be achieved by the use of reactions catalyzed by PLP-dependent enzymes (next section "Synthesis of non-canonical amino acids by PLP-dependent enzymes").

β-Hydroxy-α-amino acids

β-Hydroxy-α-amino acids (also called 3-Hydroxy-α-Amino Acids, 3-HAA;) constitute a large and widespread class of compounds. They can be found either as naturally occurring building blocks in proteins (threonine, serine and 3-hydroxyproline) and as components of many composite natural products, endowed with a wide range of biological activities such as antibiotics, immunosuppressants and peptide conjugates [27]. The non-canonical amino acids belonging to this class are useful building blocks in synthetic, combinatorial (e.g., for constructing libraries of β-lactams for antibiotic synthesis), and medicinal chemistry. Following is a list of some of the principal β-hydroxyamino acids found in nature (Fig. 7):

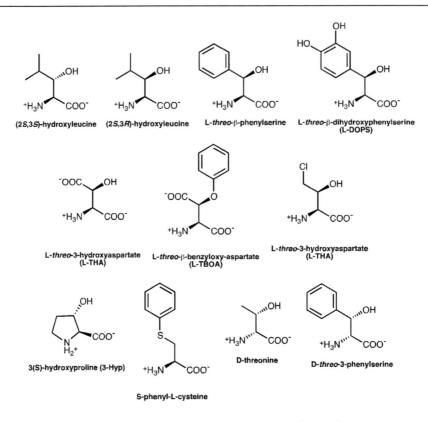

Figure 7. Structures of naturally occurring non-canonical amino acids.

- **3(S)-hydroxy-L-leucine** ((2S,3S)-hydroxyleucine), is a key component of several cyclo-depsipeptides, natural peptidic antibiotics isolated from bacteria (*Streptomyces*), fungi and marine sponges, which are often used as lead compounds for pharmacologically more potent and toxicologically safer derivatives. Cyclodepsipeptides containing 3(S)-hydroxy-L-leucine include antibiotics telomycin and variapeptin, the immunosuppressant with anti-inflammatory activity L-156,602, and several potent antitumor agents such as axinothricin, citropeptin, A83586C, and verucopeptin [28].

- **3(R)-hydroxy-L-leucine.** The configuration around the Cβ seems to be important for biological activity. As a matter of fact, the (2S,3R) isomer of hydroxyleucine is also a specific constituent of bioactive compounds, such as lactacystin and omuralide, protea-some inhibitors with neurotrophic properties [29], and lysobactins (or katanosins), a class of macrocyclic peptide lactone (depsipeptide) antibiotics also containing other proteino-genic and nonproteinogenic hydroxylated amino acids (HyPhe, HyAsn, L-*allo*-Thr), together with D-amino acids [30].

- **3-hydroxy-L-lysine**, is a naturally occurring amino acid and a putative intermediate in the synthesis of balanol, a potent protein kinase C inhibitor [31]. Its isomer 4-hydroxylisine is, on the other hand, widely known as a component of collagen.

- Besides being present as building block of antibiotics such as lysobactins [32], L-*threo*-β-**phenylserine** and its ester derivatives possess antiviral activity by themselves. This biological activity is competitively reversed by phenylalanine and abolished by the substitution of the -OH or α-amino group of phenylserine [33]. Moreover, L-*threo*-β-phenylserine was shown to act as an auxin antagonists and to possess antifungal chemotherapeutic activity [34]. An analogue of β-phenylserine, L-*threo*-**dihydroxyphenyl-serine** (L-DOPS; Droxidopa) is a psychoactive drug and synthetic amino acid precursor, which acts as a prodrug to the neurotransmitters norepinephrine (noradrenaline) and epinephrine (adrenaline) [35]. Unlike norepinephrine and epinephrine themselves, L-DOPS is capable of crossing the blood-brain barrier.

- **3-hydroxy-N^ε,N^ε,N^ε-trimethyl-L-lysine** is an intermediate of L-carnitine biosynthesis. L-carnitine is an essential molecule for fatty acid entry into the mitochondria and energy metabolism. In humans, L-carnitine is obtained either form the diet or from a four step biosynthetic pathway starting from N^ε,N^ε,N^ε-trimethyl-L-lysine. The aldolase responsible for the cleavage of the 3-hydroxy-N^ε,N^ε,N^ε-trimethyl-L-lysine in carnitine biosynthesis has not been yet identified. Cytosolic SHMT, the key metabolic PLP-dependent enzyme that typically catalyzes the interconversion of L-serine to glycine with the formation of 5,10-methylenetetrahydrofolate, was shown to catalyze *in vitro* the above mentioned reaction [36] and might be responsible for the reaction *in vivo* [37]. L-threonine aldolase, the enzyme strictly related to SHMT (see par. 3), might be responsible for the reaction in yeast [38].

- DL-*threo*-3-hydroxyaspartate (DL-**THA**) derivatives are important pharmacological agents, as they function as non-transportable inhibitors of the high-affinity Na^+-dependent excitatory amino acid transporters (EAATs), that maintain glutamate concentration below the excitotoxic level. L-glutamate is the major excitatory mediator in mammalian central nervous system and its extracellular concentration has to be tightly regulated. In particular, DL-THA derivatives possessing an ethereal bulky substituent (i. e. a benzyl or naphtyl group), are able to exert their activity on all EAAT subtypes [39].

- **4-chloro-L-threonine**, is a component of several nonribosomally synthesized phytotoxic and antifungal lipodepsinonapeptides such as syringomycin, syringotoxin, syringostatin, pseudomycin and cormycin produced by many strains of the plant colonizing bacterium *Pseudomonas spp.* [40]. Chlorothreonine is also present in two structural variants of the antitumor antibiotic compound actinomycin Z from *Streptomyces fradiae*, namely actinomycin Z_3 and Z_5. The presence of chlorine contributes to the biological activities of these natural compounds. 4-Chlorothreonine was also isolated as a free amino acid in *Streptomyces* cultures and was reported to inhibit the growth of radish, sorghum and also *Candida albicans* [41]. 4-Chlorothreonine has also been investigated as a potential antitumor agent, as a result of its ability to inactivate PLP-dependent enzyme serine hydroxymethyltransferase through a mechanism-based (suicide) inhibition. Its fluorinated analogues (such as 4-trifluoro-L-threonine or the corresponding L-*allo*-threonine derivatives) would likewise be expected to react with serine hydroxymethyltransferase and might thus be suitable for chemotherapy [42]. Another non-canonical hydroxylated amino acid, 4-hydroxythreonine, can be found in other actinomycins complexes (i.e. actinomycin Z_1) [43]. Interestingly, the phosphorylated form of this amino acid, 4-phospho hydroxy-L-threonine, is 4-phospho-hydroxy-L-threonine is an obligatory intermediate in PLP coenzyme biosynthesis in the γ-subdivision of eubacteria (such as *Escherichia coli*) [44].

- **3(S)-Hydroxy-L-proline (3-Hyp)**. First isolated in 1902 by German chemist Hermann E. Fischer from hydrolyzed gelatin, 3-hydroxy-L-proline is a common non-proteinogenic amino acid found in animal tissues. Together with its isomer 4(R)-hydroxy-L-proline, it is one of the essential constituents of collagen and related proteins such as elastin and bacterial surface proteins. The hydroxylation of proline residues highly increases the stability of the collagen triple helix, primarily through stereoelectronic effects [45]. Hydroxyprolines account for roughly 4% of all amino acids in animal tissues. Hyp is found in few proteins other than collagen, for example in hydroxyproline-rich glycoproteins (HRGP) a super-family of extracellular structural proteins found in plant cell walls, with physiological importance in cell signaling and embryogenesis and exploited for protein scaffold design for practical applications in bionanotechnology and medicine [46]. Also, Hyp is found in some snail poisons cyclopeptides, lacking collagen-like sequences [47]. Hydroxyproline is naturally generated as a post-translational modification in a complex reaction requiring oxygen, L-ascorbic acid and α-ketoglutarate and catalyzed by prolylhydroxylase. Other hydroxyprolines (such as 2,3-cis-, 3,4-trans-dihydroxyproline) are also found in nature, as part of repeated sequences in adhesive proteins from mussels, diatom cell walls and fungus poisons [48 – 50]. 3-Hyp analogue **3-hydroxy-5-methylproline** is present as a component of antibiotic actnomycin Z_1 [51].

β-Substituted amino acids

The most important derivatives of canonical sulfur-containing amino acids L-cysteine and L-methionine are their β-substituted seleno-derivatives. Selenium is a trace element essential for normal physiological processes. It is toxic at relatively low levels, and selenium compounds must be carefully administered, due to the narrowness between physiologically required dosage and toxic quantity. Selenium-containing amino acids, such as selenocysteine (Sec), selenocystine and selenomethionine (Sem) can act directly as antioxidants or chelators of redox-active metal ions, or can be incorporated into selenium-dependent antioxidant enzymes (e.g. glutathione peroxidase, thioredoxin reductase, methionine sulphoxide reductase, formate dehydrogenase, glycine reductase, deiodinase and several hydrogenases). Sec is incorporated into proteins at specific positions and participates to enzyme catalysis, enhancing the kinetic properties. For this reason it has been considered as the 21st proteinogenic amino acid. The biosynthesis of Sec is quite peculiar, in that it happens directly on its tRNA which is then recognized by the ribosome machinery and inserted through a UGA codon-directed cotranslation (mediated by the so-called SElenoCysteine Insertion Sequences (SECIS) elements and involving four specific genes, *selA-D*). The product of *selA*, selenocysteine synthase, is the PLP-dependent enzyme responsible for the last biosynthetic step. Its reaction will be described in par. 1. On the other hand, Sem is randomly incorporated into proteins in place of methionine, because tRNA^{Met} does not discriminate between Met and Sem, and seem to have no effect on protein function. Laboratory-induced incorporation of Sem into proteins allows for the production of protein crystals suitable for three-dimensional structure determination via multiwavelength anomalous dispersion phasing methods. Higher animals are not able to *de novo* synthesizes Sem and must acquire it from the diet. Selenomethionine is then incorporated into proteins and stored in the body, from which is then released by normal metabolic processes.

The unnatural derivative of L-cysteine, **S-phenyl-L-cysteine** (Fig. 7) is a building block of the anti-AIDS drug Nelfinavir (brand name Viracept), an orally bioavailable inhibitor of HIV-1 and HIV-2 proteases [52] and in phase I clinical trials as anti-cancer agent [53].

D-Amino acids

Although not used in ribosomal protein synthesis, D-amino acids are important building molecules for bacterial wall construction (the oligopeptide moiety of peptidoglycan contains D-amino acids in both Gram-positive and Gram-negative bacteria), are found in skin antimicrobial peptides [54] and are used by fungi for the biosynthesis of several antibiotics and ionophores.

D-alanine, present in bacterial peptidoglycan, is produced from the L- enantiomer by alanine racemase, one of the most investigated bacterial PLP-dependent enzyme. A distinct alanine racemase exists in eukaryotic organisms, such as the fungi *Tolypocladium niveum* and *Cochliobolus carbonum*. This enzyme has a completely different protein fold and evolutionary origin with respect to its bacterial homonym. In *T. niveum* it is essential to provide the D-alanine incorporated by non-ribosomal synthesis into the cyclic undecapeptide cyclosporin, a potent immunosuppressant drug. In *C. carbonum*, D-alanine is a component of the cyclic tetrapeptide HC-toxin, an inhibitor of histone deacetylase which is an essential virulence determinant [26].

Gramicidin is a polypeptide with alternating L- and D-amino acids, with the general formula: formyl-L-X/Gly/L-Ala/D-Leu/L-Ala/D-Val/L-Val/D-Val/L-Trp/D-Leu/L-Y/D-Leu/L-Trp/D-Leu/L-Trp-ethanolamine (X and Y being markers of different gramicidin subtypes). The alternating D-/L-amino acid sequence is essential for the formation of the characteristic β-helix, which assembles in the lipid bilayers to form pores and increase bacterial membrane permeability. The presence of D-amino acid is also essential to confer resistance to proteolitic cleavage. Interestingly, all D-enantiomer forms of naturally occurring antibiotics where shown to be completely resistant to enzymatic digestion while retained antibacterial activity [55].

D-amino acids are also widely distributed in vertebrate tissues and body fluids. For example, **D-aspartic acid** is found during the development of brain and peripheral organs on early stages of life but regulates also adult neurogenesis [56]. D-**Serine** is the most biologically active D-amino acid described to date. In the brain, it acts as a potent agonist at the glycine of the NMDA-type glutamate receptor (N-methyl-D-aspartate receptor), dramatically increasing receptor affinity for glutamate; D-serine is then able to produce excitotoxicity, without any change in glutamate concentration *per se* [57]. High concentration of D-amino acids in human serum, such as **D-Ala** (but also D-Ser), are suggested to correlate with damage to renal function [58].

Several enzymes acting on a variety of **3-hydroxy-D-amino acids** have been described and characterized from different microorganisms. These enzymes are highly specific for the D-amino acid forms. One example is D-*threo*-**3-phenylserine/D-threonine** (Fig. 7) dehydrogenase from *Pseudomonas cruciviae*, that catalyzes the oxidation of the 3-hydroxyl group of to yield D-2-amino-3-ketobutyrate, which is spontaneously decarboxylated into aminoacetone [59]. D-Threonine and D-*threo*-3-phenylserine are synthetic compounds and have not been found in the free form in nature, although the toxic peptide phalloidine contains D-threonine [60]. Another example is D-3-hydroxyaspartate aldolase from *Paracoccus denitrificans*, a PLP-dependent enzyme which catalyzes the conversion of D-**3-hydroxyaspartate** to glyoxylate and glycine. This enzyme is strictly specific for the D-amino acid form, but does not distinguish between the *threo*/*erythro* for at β-carbon. It also acts on D-threonine, D-3-phenylserine and on norepinephrine analogues D-3 – 3,4-dihydroxyphenylserine and D-3 – 3,4-methylenedioxyphenylserine [61]. D-hydroxyaspartate is found in phallacidin, the major *Amanita* mushroom toxin [62]. The physiological role of these D-hydroxyamino acids (and of the enzymes acting on them), is not clear, but could be related to the fungus *vs.* bacteria host-pathogen interaction.

Biogenic amines

Biogenic amines are important compounds found in microbial, vegetable and animal cells. They are mainly produced via amino acid decarboxylation although some of the aliphatic amines can be formed *in vivo* by amination or transamination of the corresponding aldehydes and ketones. Besides their role as nitrogen source for the biosynthesis of hormones, alkaloids, nucleic acids and proteins, biogenic amines are precursors of food aroma components and can be related to food spoilage and fermentation processes. In addition, they can be precursors of carcinogenic *N*-nitroso compounds. The biological function of biogenic amines in humans is of paramount importance in the central nervous system. A list of common amino acid-derived amines acting as neurotransmitter and hormones is shown in Table 1 (and Fig. 8).

Table 1. Biogenic amines derived from amino acids and their biological effects.

BIOGENIC AMINE	PRECURSOR AMINO ACID	BIOLOGICAL ACTIVITY	
Histamine	Histidine	Neurotransmitter Pro-inflammatory signal from mast cells Mediator	Mediates arousal/attention Allergic reactions/tissue damage Stimulation of HC 1 secretion in the stomach. Lowers blood pressure
Serotonin	Tryptophan	CNS neurotransmitter	Regulates mood, appetite, sleep
Norepinephrine (noradrenaline) **Epinephrine** (adrenaline) **Dopamine**	Tyrosine	Neurotransmitter Hormone Neurotransmitter	Involved in wakefulness and sleep Regulates heart rate, blood vessel and air passage diameters. Fight-or-flight response, stress hormone Important roles in behavior and cognition, voluntary movement, motivation, punishment and reward, sleep, mood, attention, working memory, learning.
Tyramine/ Phenylethylamine	Tyrosine	Mediator	Rise of blood pressure. Catecholamine releasing agent.
3-Iodothyronamine	Thyroid hormones	Hormone	Decreased body temperature and cardiac output
Tryptamine and its derivatives	Tryptophan	Neuromodulators	Modulator of serotonin effects on CNS. Functions by itself in central neurotransmission still debated

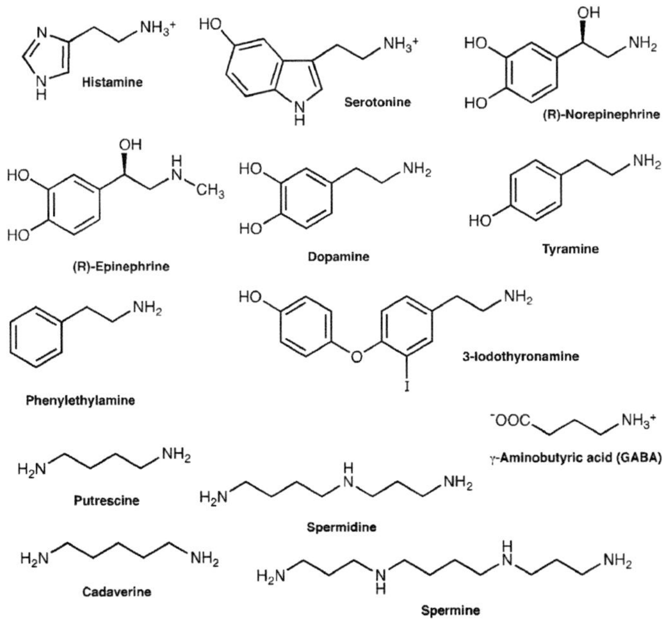

Figure 8. Structure of physiologically relevant biogenic amines.

Polyamines such as cadaverine, putrescine, spermidine, and spermine (Fig. 8) (derived from amino acids arginine, lysine and ornithine) are linear cationic molecules that interact with DNA and regulate nucleic acid function. They have also been shown to promote pro-grammed ribosomal frame shifting during translation [63] and to modulate membrane per-meability acting on a variety of ion channels. Although every cell is able to synthesize polyamines, the body relies on a continuous dietary supply of these compounds, stored in different organs and then release in a strictly regulated manner.

Many PLP-dependent decarboxylases are involved in the production of biologically impor-tant amines. DOPA decarboxylase catalyzes the conversion of L-3,4-diidroxyphenylalanine (L-DOPA) into dopamine, a neurotransmitter found in the nervous system and peripheral tissues of both vertebrates and invertebrates and also in plants where it is implicated in the biosynthesis of benzylisoquinoline alkaloids. It also catalyzes the decarboxylation of 5-hy-droxy-L-tryptophan to give 5-hydroxytryptamine (serotonin). Ornithine decarboxylase cata-lyzes the rate-limiting step in the biosynthesis of polyamines, i.e. the decarboxylation of the urea cycle intermediate L-ornithine to give putrescine. Glutamate decarboxylase is involved in the formation of γ-aminobutyric acid (GABA; Fig. 8), the major inhibitory neurotrans-mitter in the central nervous system.

SYNTHESIS OF NON-CANONICAL AMINO ACIDS BY PLP-DEPENDENT ENZYMES

Synthesis of β-hydroxy-α-amino acids

The synthesis of β-hydroxy-α-amino acids (3-HAAs) has been attracting a lot of attention because many of these compounds are biologically active or constitute intermediates or building blocks of drugs. The industrial production of 3-HAAs has been mainly restricted to chemical synthesis, which yields all four stereoisomers (D- and L- *threo* and *erythro* forms) and therefore requires further steps of purification. However, in the recent past many communications concerning the use of enzymes for the synthesis of 3-HAAs have appeared in the literature. Two types of fold type I PLP-dependent aldolases have been used for synthetic purposes, both catalyzing the reversible aldol reaction of glycine with an aldehyde: **serine hydroxymethyltransferase** (SHMT; systematic name *5,10-methylenetetrahydrofola-te:glycine hydroxymethyltransferase*, EC 2.1.2.1) and **threonine aldolase** (TA; systematic name *threonine acetaldehyde-lyase*, EC 4.1.2.x). 3-HAAs contain two chiral centers, one at Cα, which determines the L- or D- configuration, and the second at Cβ, responsible for the *erytro* or *threo* configuration of L- and D-3-HAAs. Although SHMT is selective for the l-3-HAA, it lacks specificity for the configuration of Cβ (SHMT from corn and rabbit liver yielded 3-HAAs with prevailing *erythro* configurations; pig liver SHMT has been used in large preparative reactions, obtaining mixtures of *erythro* and *threo* compounds [64]. For this reason, the synthesis of 3-HAAs catalyzed by SHMT has not been pursued with great passion.

On the other hand, many TAs with different stereospecificity have been isolated from a number of bacteria and fungi and characterized [65]. According to their stereospecificity at Cα, TAs are classified into L- and D- type enzymes. The former ones are further divided into low-specificity TAs (EC 4.1.2.48), L-threonine aldolases (EC 4.1.2.5) and L-*allo*-threonine aldolase (EC 4.1.2.49), depending on their preference for the configuration at Cβ. Only low specific D-TAs (EC 4.1.2.42) are known so far.

TA have been used for the synthesis of 3-HAAs following two strategies: the enzymatic resolution of racemic mixtures obtained via chemical synthesis or the direct enzymatic asymmetric synthesis.

The enzymatic resolution strategy has been extensively employed [64, 66] and, significantly, has been reported in multi-step procedures for the synthesis of the antibiotic thiamphenicol [67] and the anti-Parkinson's disease drug L-*threo*-DOPS [68]. It consists in the enantioselective cleavage of one enantiomer of a diastereomerically highly enriched racemate, previously obtained through chemical synthesis. The drawback of this strategy is that its maximum yield is 50% and therefore its attractiveness relies on the economical chemical synthesis of the racemate.

The TA-based enzymatic asymmetric synthesis of 3-HAAs is a very attractive synthetic route, which has been followed by many researchers. Several are the successful achievements in this field, although the low to medium diastereoselectivity of the biotransformations obtained still needs to be improved. Detailed research on the synthetic capabilities [69, 70] and stereospecificity [71] of TAs has been carried out. Attention has also been focused on the production of the anti-Parkinson's disease drug L-*threo*-DOPS [72] and 3-HAAs to be used in the synthesis of complex organic compounds, such as peptide mimetics [73].

The selectivity in the synthesis of either *threo* or *erythro* 3-HAAs depends on both the stereospecificity of the enzymes and on the thermodynamic equilibrium of the aldol reaction. A high *erythro* over *threo* selectivity could be obtained in the TA-catalyzed condensation of γ-benzyloxybutanal and glycine under kinetically controlled conditions [74]. The *erythro* product was used in the synthesis of mycestericin D, a potent immunosuppressant. On the other hand, mutant forms of TA were produced by error-prone PCR followed by a high-throughput screening, which displayed an increased diastereoselectivity for L-*threo*-DOPS synthesis [75, 76]. Recently, a novel genetic selection system has been devised in the attempt to expand the substrate scope and enhance the selectivity of TAs [77].

The catalytic capabilities of TAs have been also exploited using substrates other than glycine and simple aldehydes. A two-step, one-pot bioenzymatic reaction, involving TA and a PLP-dependent decarboxylase, has been devised to produce β-amino alcohols [71, 78]. L-TA from *Escherichia coli* was shown to accept aldehydes bearing a carboxylic acid, forming ω-carboxy-β-hydroxy-L-α-amino acids [39]. Moreover, the enantio- and diastereoselective synthesis of α,α-disubstituted β-hydroxy-L-α-amino acids was obtained using L-*allo*-threonine aldolase from *A. jandeii* and D-threonine aldolase from *Pseudomonas* sp. as catalysts and D-amino acids (alanine, serine and cysteine) as substrates [79]. Interestingly, an engineered D-alanine racemase was employed as an aldolase with D-alanine as substrate [80– 83]. This latter achievement supports the hypothesized strict structural and functional relationship between PLP-dependent racemases and aldolases [14].

Finally, the use of whole cells for the threonine aldolase-based synthesis of L-*threo*-DOPS by TA has been recently reported [75, 84].

β-Substitution reactions

This kind of reactions, catalyzed by several PLP-dependent enzymes, allow for the addition of different nucleophiles to an aminoacrylate intermediate previously formed from the starting substrate upon a β-elimination reaction. Subsequently, the product of condensation (the amino acid analogue) is released from the cofactor. These reactions can accomplish carbon-carbon, carbon-nitrogen or carbon-sulfur (or selenium) bond formation, depending on the nucleophile used as second substrate (indoles, phenols, cyanides for C-C bond formation; pyrroles, azides for C-N bond; thiols/selenols for C-S or C-Se bond formation). The most used PLP-dependent enzymes for β-substitution reactions and production of amino acid analogues are herein described.

L-Tyrosine phenol-lyase (TPL; EC 4.1.99.2). The synthesis of fluorinated analogues of tyrosine by use of bacterial tyrosine phenol-lyase has been described many years ago [85]. The physiological reaction catalyzed by TPL (a fold type I PLP-dependent enzyme) is the hydrolytic cleavage of L-tyrosine to yield phenol and ammonium pyruvate. In addition to this reaction, participating in tyrosine metabolism, TPL also efficiently catalyzes the β-elimination of a number of β-substituted amino acids with good leaving groups, such as S-methyl-L-cysteine, β-chloroalanine, and S-(*O*-nitrophenyl)-L-cysteine. TPL has also been shown to catalyze the racemization of alanine, but at a much slower rate [86]. The reaction of TPL is readily reversible at high concentration of ammonia and pyruvate. Under these conditions the enzyme binds ammonia first, followed by pyruvate and then phenol. The reverse reaction catalyzed by TPL can be used then as a biosynthetic tool for the production of tyrosine and tyrosine analogues. When an appropriate phenol derivative is substituted for phenol, for example, the corresponding tyrosine analogues can be synthesized. Several differently substituted fluoro-, chloro-, bromo-, iodo-, methyl-, methoxy-L-tyrosines have been produced by this method. If the phenol ring is substitute by a catechol, 3,4-dihydroxy-phenil-L-alanine (L-DOPA) is produced. L-DOPA is used in the treatment of Parkinson's disease and more than 250 tons of it is made every year for pharmaceutical application. Most of it is indeed produced by an industrialized process that use engineered microorganisms overexpressing TPL activity [87].

Substantially homologous to TPL in sequence, three-dimensional structure and reaction mechanism is **tryptophanase** (Tnase or Trpase; systematic name *L-tryptophan indole-lyase*, EC 4.1.99.1) a bacterial PLP-dependent lyase that catalyzes in vivo degradation of L-tryptophan to yield indole, pyruvate and ammonia [88]. Also in this case, the enzyme can act in the reverse direction, to synthesize L-tryptophan in conditions of excess pyruvate, ammonia and a supply of indole. The enzyme also catalyzes α,β-elimination and β-replacement reactions on several other β-substituted L-amino acids. To our knowledge, this enzyme has not been exploited for biotechnology applications.

Tryptophan synthase (TS; systematic name L-*serine hydro-lyase*, EC 4.2.1.20.) can be used for the synthesis of tryptophan analogues. Tryptophan synthase catalyzes, in bacteria fungi and plants, the final two steps in tryptophan biosynthesis. The enzyme is a $\alpha_2\beta_2$ tetramer. The physiological reaction of TS is the combination of the reactions occurring at the α and β-sites, which are tightly coupled through allosteric interactions. The reaction catalyzed by the α-subunit is the reversible retroaldol cleavage of indole-3-glycerol phosphate to give indole and D-glyceraldehyde-3-phosphate. The indole is not released into the solvent, but is directly transferred to the β-subunit (a fold type II PLP-enzyme) where it is condensed to the aminoacrylate intermediate formed by the elimination of water from the L-serine-PLP adduct. The protonation of the α-carbon of the following quinonoid intermediate produces the external aldimine with L-tryptophan ([24, 89] and references within).

The α-subunit reaction is catalyzed by general acid-base catalysis; although it is readily reversible and thus could be used for the synthesis of indole-3-glycerol phosphate and its derivatives (through indole and aldehydes analogues), there have been very few investigations about its synthetic applications. On the other hand, the β-reaction of TS has been used for the preparation of a large variety of non-canonical α-L-amino acids. In fact, a wide range of indole analogues are recognized by the β-subunit active site and give rise to the corresponding L-tryptophan derivatives. Substituted methyl-, fluoro-, chloro-, hydroxy-, methoxy-, dipluoromrthyl-, azido-indoles in various positions have been proved to be accepted as substrates by tryptophan synthases from several organisms (*Neurospora crassa, E. coli, S. typhimurium*) although with catalytic constants different from unsubstituted indole. For example, the much lower nucleophilicity of azaindoles requires the reaction for the formation of azatryptophans to be carried out for longer times (even days or weeks). TS is also able to accommodate and condense heterocyclic double 5-membered rings such as thienopyrroles and selenopyrroles with the formation of thiatryptophans and selenatryptophans, whose preparation by synthetic chemistry would prove difficult due to their extreme acid sensitivity; these compounds, incorporated into protein, may be useful as spectroscopic probes and heavy atom derivatives in protein crystallography [90 – 92]. Notably, reaction with all indole derivatives take place with high stereospecificity, since the reprotonation of $C\alpha$ only happens on the *re*-face of the quinonoid intermediate.

***O*-acetylserine sulfhydrylase** (OASS, cysteine synthase; systematic name *O3-acetyl-L-serine:hydrogen-sulfide 2-amino-2-carboxyethyltransferase*, former EC 4.2.99.8, now 2.5.1.47) the enzyme responsible of the final step of cysteine biosynthetic pathway in bacteria, fungi, plants and protozoan parasites can be used for the production of a variety of β-substituted amino acids. This enzyme catalyzes a β-replacement reaction in which the β-acetoxy group of *O*-acetyl-L-serine (OAS) is replaced by bisulfide to give L-cysteine and acetate. The substrate *O*-acetyl-L-serine is synthetized by serine acetyltransferase (SAT), from L-serine and acetyl-CoA. Interestingly, SAT modulates OASS activity by directly interacting with it. The reaction follows a ping-pong bi-bi mechanism with a stable α-amino-acrylate intermediate. The enzyme is a homodimer belonging to fold type II PLP-dependent enzymes, the active site residues being contributed by a single subunit [93]. Given the key biological role played by *O*-acetylserine sulfhydrylase in bacteria, inhibitors with potential antibiotic activity have been developed [94]. Bacteria such as *E. coli*, *S. typhimurium* and *Haemophilus influenzae* possess two OASS isoforms (namely OASS-A and OASS-B, encoded by *cysK* and *cysM* genes, respectively). The two enzymes share a 43 % sequence identity, have almost superimposable structure, and similar activities. Whereas OASS-A is highly expressed at basal level, the physiological role of OASS-B is still controversial. Both enzymes have relaxed substrate specificity, with OASS-B being more promiscuous and able to accept bigger substrates. A recent study on both isoforms of *E. coli* OASS showed that many different nucleophiles can replace bisulfide in the reaction catalyzed by OASS, such as several types of thiols (methane-, ethane-, propylene-, phenyl-thiol) and selenols, phenol, azide, cyanide, 5-memberd rings containing at least two adjacent nitrogens (i. e. pyrazole, but not imidazole), giving rise to a variety of non-canonical β-substituted-L-α-amino acids [95, 96]. OASS-B isoform was able to synthetize, starting from *O*-acetyl-L-serine and 1,2,4-oxadiazolidine-3,5-dione quisqualic acid, an amino acid extracted from plants with neuro-stimolatory effects by acting as an agonist of AMPA and metabotropic glutamate receptors. Interestingly, its decarboxylated counterpart, quisqualamine, shows central depressant and neuroprotective properties and appears to act predominantly as an agonist of the GABA$_A$ (γ-aminobutyric acid, the decarboxylated form of glutamate) receptor.

Similar to OASS is ***O*-phosphoserine sulfhydrylase** (OPSS; systematic name *O-phospho-L-serine:hydrogen-sulfide 2-amino-2-carboxyethyltransferase*, EC 2.5.1.65), an enzyme recently discovered in hypothermophilic aerobic archea which catalyzes the sulfhydrylation of *O*-phospho-L-serine to form L-cysteine. *O*-phosphoserine is much more stable than *O*-acetylserine and might be the precursor involved in cysteine biosynthesis in extremophiles. As for OASS, OPSS is able to catalyze synthetic reactions using a variety of nucleophiles, giving rise to unnatural amino acids. This enzyme was actually also found in organisms such as the human parasite *Trichomonas vaginalis* and pathogens *Mycobacterium tuberculosis*, revealing a new cysteine biosynthetic pathway, with interesting implication for the design of innovative drugs.

Another enzyme closely related to OASS is **O-acetyl-L-homoserine sulfhydrylase (OAHS**; homocysteine synthase, *O*-acetylhomoserine(thiol)-lyase; EC 4.2.99.10). This is a sulfide-utilizing enzyme involved in the complex and diversified L-cysteine and L-methionine biosynthetic pathways of various bacteria, filamentous fungi and yeast [97]. OAHS is known to catalyze the conversion of *O*-acetyl-L-homoserine to L-homocysteine using bisulfide as direct sulphur donor. L-Homocysteine is then converted to L-methionine through tetrahydro-folate dependent methylation catalyzed by methionine synthase. OAHS is essential in certain microorganisms, whereas others use alternative methionine biosynthetic pathways. Remark-ably, OAHS is structurally related to cystathionine α-, β-, γ-lyases from bacteria, plants, yeast and animals, all enzymes involved in methionine biosynthesis and requiring PLP as cofactor. The reaction catalyzed by OAHS is a γ-substitution reaction. The first half reaction requires two deprotonation steps, followed by γ-elimination of the acetyl group, leaving an aminoacrylate intermediate. Bisulfide then makes a nucleophilic attack to form homocys-teine. As for OASS, different nucleophiles can perform this second step of the reaction, including Na_2Se_2 [98] cyanide, sodium azide, methanethiol (directly generating methionine), ethanethiol and thiophenol [99]. Recent studies indicated that in *Wolinella succinogenes*, a close relative of *Helicobacter pylori*, the source of sulphur in methionine biosynthesis is a protein thiocarboxylate. The substrate of OAHS, in this case, would then not be a small nucleophile, but an entire sulphur-carrier low molecular weight protein (with ubiquitin-like fold) [100].

Finally, to conclude this general view on PLP-dependent enzymes catalysing β-substitution reactions, it is worth mentioning the enzymes that naturally produce seleno-amino acids. The reaction mechanism of **selenocysteine synthase** (systematic name *selenophosphate:O-phos-pho-L-seryl-tRNASec selenium transferase*, EC 2.9.1.2), the enzyme responsible of the last step of selenocysteine biosynthesis in eukaryotes and Archea is very similar to the one catalyzed by OPSS. Interestingly, in this case the substrate is not free *O*-phospho-L-serine, but *O*-phosphoseryl-tRNA[Ser]Sec, which is generated by enzymatic phosphorylation of seryl-tRNA[Ser]Sec. First, the enzyme removes the phosphate group from *O*-phosphoseryl moiety to yield the aminoacrylate intermediate, and then accepts activated monoseleniumphosphate as condensing nucleophile. The bacterial (*E. coli*) counterpart, L-**Seryl-tRNASec selenium transferase** (systematic name *Selenophosphate:L-seryl-tRNASec selenium transferase*, EC 2.9.1.1), catalyzes the same reaction using seryl-tRNA[Ser]Sec, generating the aminoacrylate through the elimination of the hydroxyl group from the seryl moiety.

NON-CANONICAL AMINO ACIDS FOR GENETIC CODE EXPANSION

What is genetic code expansion?

The number of amino acids in ribosomal protein synthesis is restricted to the 20 canonical amino acids. Additional functionalities into protein structures can be introduced by expand-ing the repertoire of amino acids that make up the basic structure of the resulting protein molecule by re-assigning one (or several) of the codons that are normally used to encode conventional amino acids (or to terminate protein synthesis) into another, non-canonical amino acid (ncAA). By incorporating an alternative amino acid into proteins, the available chemical options to achieve additional functionalities is substantially enlarged. This approach relies on

1. the availability of a tRNAs/amino acyl-tRNA synthetase pairs that allow to "mischarge" the tRNA of the re-assigned codon with the ncAA and

2. a mechanism that prevents the re-assigned codon to fulfill its original function with a canonical amino acid or a termination function (STOP codon).

The whole system is defined to consist of four elements: a re-assigned codon, a cognate mischarged tRNA, a cognate mischarging amino acyl-tRNA synthetase, and the charged ncAA. Through this approach, the biological, chemical, or physical properties of new amino acids are precisely defined by the chemist at the bench and, due to the genetic encoding of these ncAAs, their incorporation into proteins should occur with exquisite fidelity and efficiency (for review see: [101]).

In this way, microorganisms with an engineered genetic code are capable of delivering the biological, chemical, or physical properties of many unnatural or synthetic non-canonical amino acids, into resulting polypeptide sequences. For example, ncAAs leading to changes in the protein backbones, such as α-hyroxy acids, can be site-specifically installed into proteins to define specific sites for chemical hydrolysis [102]. Other examples are special ncAAs building blocks such as fluorinated analogues that have highly attractive unique features for pharmaceutical industry, since fluorination improves bioavailability (i. e. membrane interaction and passage activity) of drugs and their metabolic stability [103]. In addition, ncAAs can be used to site-specifically derivatize proteins with PEG molecules, sugars, oligonucleotides, fluorophores, peptides, and other unique synthetic moieties. Important bioorthogonal[2] chemical functionalities such as azides, olefins, carbonyl compounds (ketones/aldehydes), strained- and unstrained-alkynes, halogens, boronic esters/acids, oximes/hydrazones can be used for specific coupling with these moieties.

Bioorthogonal click-reactions for site-specific chemoselective ligations include copper-click, photo-click, metathesis, and catalyzed oxime/hydrazone chemistries. Remarkably, these chemistries are often mutually orthogonal: e. g., photoclick chemistry is bioorthogonal to both oxime/hydrazone and copper-click chemistry, and should enable a variety of different coupling reactions to the same protein in a single expression experiment (for review see [104]).

---

[2] The term bioorthogonal chemistry refers to any chemical reaction that can occur inside a living system without interfering with native biochemical processes.

While *in vitro* incorporation of ncAAs has been known for decades, *in vivo* approaches that are currently available for the incorporation of desirable ncAAs in a controlled manner have only been recently developed and can be divided into two methodologies. In the first approach, residue-specific incorporation of different ncAAs into target proteins occurs via sense codon reassignment using auxotrophic host strain that exploits wide substrate tolerance in activating similar amino acids (analogs, surrogates) [105]; the second approach includes suppression methodologies and use orthogonal pairs to reassign termination or non-triplet coding units for site-specific addition of ncAAs to the existing amino acid repertoire during translation. In this case, aminoacyl-tRNA synthetases needed to activate the desired ncAAs are selected using standard positive/negative selection methods [106, 107].

Code expansion and metabolic engineering: crucial role of PLP-dependent enzymes

One of the most straightforward methods, using expanded genetic code to generate unnatural polypeptides with non-canonical amino acids usually require the addition of the starting building blocks to the growth medium and their subsequent uptake by the cellular transport machinery. While this is perfectly acceptable for small-scale experiments, there is still no solution for mass production. Here, the ncAAs should be produced from medium nutrients, just as all other cell components, to prevent complicated feeding schemes and additional costly substrates. To tackle this general problem it is necessary to perform intracellular synthetic pathway engineering. Thereby, PLP-dependent enzymes could be of prime importance, as they are well known to be catalytically promiscuous and to contain members from thermophilic group of bacteria, which provide a rather stable scaffold for bioengineering purposes [108].

Certainly, one of the best documented semisynthetic approach of non-canonical amino acids synthesis through metabolic engineering using PLP-dependent enzymes is the engineering of cysteine biosynthetic pathway as reported by Maier [95]. In particular, the final step in this biosynthetic pathway is catalyzed by *O*-acetylserine sulfhydrylase, a PLP-dependent enzyme which catalyzes β-substitution reaction on *O*-acetylserine. The catalytic promiscuity of this enzyme is reflected in its broad substrate specificity in a similar manner as tryptophan synthase (see above). Combined with this feature, the intracellular deregulation of cysteine pathway enables the biosynthesis of amino acid derivatives characterized by diverse side chains with interesting chemical functionalities. For example, the fermentation media supplied with toxic substances such as azide, cyanide or triazole allows their biotransformation into amino acids such as azidoalanine, cyanoalanine and triazole-1-yl-alanine (Fig. 9). Using this approach, high yield production of non-canonical amino acids are reported as opening a fairly good perspective to make a further step and couple such reengineered metabolic processes with reprogrammed protein translation apparatus for tailor-made protein production on industrial scale.

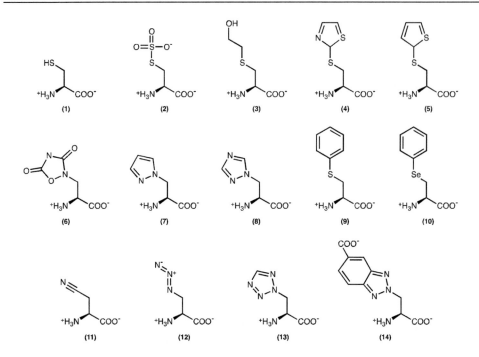

Figure 9. Structures of amino acids generated by metabolic engineering by means of the catalytic promiscuity of *O*-acetylserine sulfhydrylase, the last enzyme in cysteine biosynthetic pathway. The canonical amino acid cysteine **(1)** is the natural product of the intracellular catalytic activity of this enzyme. Non-canonical amino acids such as azidoalanine **(12)** are useful for protein surface diversification by Staudinger ligation or for 'click' chemistry of proteins; cyanoalanine **(11)** might serve as biophysical marker (IR probe); *S*-phenylcysteine **(9)** is an important building block in the design of inhibitors for AIDS therapy. Common names of the other non-canonical amino acids shown in the figure are: *S*-sulfocysteine **(2)**, *S*-hydroxyethylcysteine **(3)**, *S*-thiazole-2-yl-cysteine **(4)**, *S*-thien-2-yl-cysteine **(5)**, 1,2,4-oxadiazolidinedionyl-alanine **(6)**, pyrazole-1-yl-alanine **(7)**, triazole-1-yl-alanine **(8)**, tetrazole-2-yl-alanine **(13)**, *S*-phenyl-cysteine **(9)**, phenyl-selenocysteine **(10)** and 5-carboxybenzotriazole-2-yl-alanine **(14)**. (Adapted from [95])

The manipulation of the tryptophan biosynthetic pathway in combination with incorporation of non-canonical amino acids into proteins is another very instructive example of biosystems engineering by coupling expanded genetic code and semisynthetic/synthetic metabolism mediated by PLP-dependent enzymes. In particular, last steps in tryptophan biosynthesis include indole production and its condensation with L-serine to make L-tryptophan. This two combined reactions are performed by tryptophan synthase, through the well characterized reaction mechanism described before, which includes direct transfer of the indole intermediate between α and β subunits through a tunnel in the enzyme complex (the so-called 'channeling effect'). PLP-dependent β subunit of tryptophan synthase, which catalyzes β-substitution reaction on indole, is one of the best-studied enzymatic systems in biochemistry. One remarkable property of this enzyme, essential for amino acid derivative production, is its broad substrate specificity; even amino acids not structurally and chemically related to tryptophan can be synthesized by using this enzyme [109] (but, significantly, these related amino acids are not substrates for endogenous tryptophanyl-tRNA synthetase). Thus, in a fermentation medium provided with a variety of (natural or synthetic) indole analogues/surrogates and controlled expression system, related amino acids analogues can be synthesized intracellularly and subsequently incorporated into target proteins as recently demonstrated [110].

From a biotechnological point of view, the great advantage of an expanded genetic code efficiently coupled with engineered PLP-dependent enzymes would be the possibility to achieve high chemical diversity at low genetic cost and to avoid supply of expensive precursors. In this way, a pathway for the biosynthesis of desired non-canonical amino acids can be engineered, imported and integrated into cellular metabolism enabling microbial hosts to generate desired amino acid from simple precursors or carbon sources.

OUTLOOK

Chemical synthesis of desired non-canonical amino acids is usually expensive. Moreover, a main drawback of chemical syntheses of ncAAs as substrates for ribosomal polypeptides synthesis, with only a few exceptions, is that the substance of interest has to be added to the growth medium and subsequently taken up by the cellular uptake machinery. This can be suitable for small laboratory-scale experiments; however, it is highly unsuitable for large-scale fed-batch production. Therefore, it is highly desirable that synthetic amino acids of interest are produced by the host from medium nutrients, i.e. from simple and economically-favorable substrates without any complicated feeding procedures. Thus, in addition to expanding the scope of protein synthesis, one of the biggest challenge in the field will be to engineer metabolic pathways so as to provide cells with target synthetic amino acids produced intracellularly from simple carbon sources or precursors (Fig. 10). In this respect, enzymatic pathways mediated by PLP-dependent enzymes are most promising targets.

Budisa, N. *et al.*

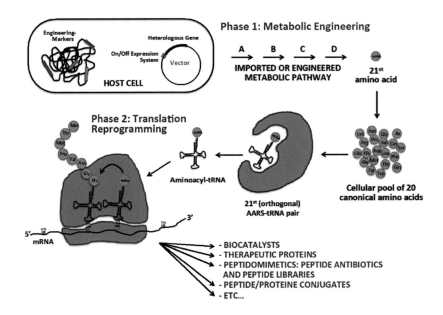

Figure 10. Coupling metabolic engineering with re-programmed protein translation. Combined together, natural products chemistry and genetic code engineering would lead to a larger scale screening for suitable amino acids as candidates for entry into the genetic code from intermediary metabolism of various species. The natural biosynthetic pathways can be imported and integrated into the metabolism of host cells in order to generate intracellular production of desired non-canonical amino acids (which should be exclusive substrates for AARS, aminoacyl-tRNA synthatase). These pathways can be attached to the existing ones in host cells, further modified, reduced, extended or optimized for balanced synthesis of desired substrates. The ultimate goal is the *in-vivo* evolution of novel synthetic pathways capable to generate substrate diversity to an extent far beyond natural one (adapted form [111]).

REFERENCES

[1] Percudani, R., and Peracchi, A. (2003) A genomic overview of pyridoxal-phosphate-dependent enzymes. *EMBO reports* **4**:850 – 854.
doi: 10.1038/sj.embor.embor914.

[2] John, R.A. (1995) Pyridoxal phosphate-dependent enzymes. *Biochim. Biophys. Acta* **1248**:81 – 96.
doi: 10.1016/0167-4838(95)00025-P.

[3] Eliot, A.C., and Kirsch, J.F. (2004) Pyridoxal phosphate enzymes: mechanistic, structural, and evolutionary considerations. *Annual review of biochemistry* **73**:383 – 415.
doi: 10.1146/annurev.biochem.73.011303.074021.

[4] Amadasi, A., Bertoldi, M., Contestabile, R., Bettati, S., Cellini, B., di Salvo, M.L., Borri-Voltattorni, C., Bossa, F., and Mozzarelli, A. (2007) Pyridoxal 5'-phosphate enzymes as targets for therapeutic agents. *Current medicinal chemistry* **14**:1291 – 1324.
doi: 10.2174/092986707780597899.

[5] Metzler, D.E., Ikawa, M., Snell, E.E. (1954) A General Mechanism for Vitamin B6-catalyzed Reactions. *J. Am. Chem. Soc.* **76**(3):648 – 652.
doi: 10.1021/ja01632a004.

[6] Dunathan, H. C. (1966) Conformation and reaction specificity in pyridoxal phosphate enzymes. *Proceedings of the National Academy of Sciences of the U.S.A.* **55**:712 – 716.
doi: 10.1073/pnas.55.4.712.

[7] Olivard, J., Metzler, D.E., and Snell, E.E. (1952) Catalytic racemization of amino acids by pyridoxal and metal salts *J. Biol. Chem.* **199**:669 – 674.

[8] Metzler, D.E., and Snell, E.E. (1952) Deamination of serine. I. Catalytic deamination of serine and cysteine by pyridoxal and metal salts. *J. Biol. Chem.* **198**:353 – 361.

[9] Denesyuk, A.I., Denessiouk, K.A., Korpela, T., and Johnson, M.S. (2002) Functional attributes of the phosphate group binding cup of pyridoxal phosphate-dependent enzymes. *Journal of molecular biology* **316**:155 – 172.
doi: 10.1006/jmbi.2001.5310.

[10] Spenser, I.D., and Hill, R.E. (1995) The biosynthesis of pyridoxine. *Natural product reports* **12**:555 – 565.
doi: 10.1039/np9951200555.

[11] Surtees, R., Mills, P. and Clayton, P. (2006) Inborn errors affecting vitamin B6 metabolism, *Future neurol.* 1F(5):615 – 620.
doi: 10.2217/14796708.1.5.615.

[12] di Salvo, M.L., Safo, M.K., and Contestabile, R. (2012) Biomedical aspects of pyridoxal 5'-phosphate availability. *Front. Biosci. (Elite Ed)* **4**:897 – 913.

[13] Mehta, P.K., and Christen, P. (2000) The molecular evolution of pyridoxal-5'-phosphate-dependent enzymes. *Advances in enzymology and related areas of molecular biology* **74**:129 – 184.

[14] Paiardini, A., Contestabile, R., D'Aguanno, S., Pascarella, S., and Bossa, F. (2003) Threonine aldolase and alanine racemase: novel examples of convergent evolution in the superfamily of vitamin B6-dependent enzymes. *Biochim. Biophysi. Acta* **1647**:214–219.
doi: 10.1016/S1570-9639(03)00050-5.

[15] Nam, H., Lewis, N.E., Lerman, J.A., Lee, D.H., Chang, R.L., Kim, D., and Palsson, B.O. (2012) Network context and selection in the evolution to enzyme specificity. *Science* **337**:1101–1104.
doi: 10.1126/science.1216861.

[16] O'Brien, P.J., and Herschlag, D. (1999) Catalytic promiscuity and the evolution of new enzymatic activities. *Chemistry & biology* **6**:R91-R105.
doi: 10.1016/S1074-5521(99)80033-7.

[17] Khersonsky, O., and Tawfik, D.S. (2010) Enzyme promiscuity: a mechanistic and evolutionary perspective. *Annual reviews of biochemistry* **79**:471–505.
doi: 10.1146/annurev-biochem-030409-143718.

[18] Kochhar, S., and Christen, P. (1992) Mechanism of racemization of amino acids by aspartate aminotransferase. *Eur. J. Biochem.* **203**:563–569.
doi: 10.1111/j.1432-1033.1992.tb16584.x.

[19] Kochhar, S., and Christen, P. (1988) The enantiomeric error frequency of aspartate aminotransferase. *Eur. J. Biochem.* **175**:433–438.
doi: 10.1111/j.1432-1033.1988.tb14213.x.

[20] Soper, T.S., Ueno, H., and Manning, J.M. (1985) Substrate-induced changes in sulfhydryl reactivity of bacterial D-amino acid transaminase. *Archives of Biochemistry and Biophysics* **240**:1–8.
doi: 10.1016/0003-9861(85)90001-3.

[21] John, R.A., and Fasella, P. (1969) The reaction of L-serine O-sulfate with aspartate aminotransferase. *Biochemistry* **8**:4477–4482.
doi: 10.1021/bi00839a038.

[22] John, R.A., and Tudball, N. (1972) Evidence for induced fit of a pseudo-substrate of aspartate aminotransferase. *Eur. J. Biochem.* **31**:135–138.
doi: 10.1111/j.1432-1033.1972.tb02510.x.

[23] Contestabile, R., and John, R.A. (1996) The mechanism of high-yielding chiral syntheses catalysed by wild-type and mutant forms of aspartate aminotransferase. *Eur. J. Biochem.* **240**:150–155.
doi: 10.1111/j.1432-1033.1996.0150h.x.

[24] Miles, E.W. (1991) Structural basis for catalysis by tryptophan synthase. *Advances in enzymology and related areas of molecular biology* **64**:93 – 172.

[25] Contestabile, R., Paiardini, A., Pascarella, S., di Salvo, M.L., D'Aguanno, S., and Bossa, F. (2001) L-Threonine aldolase, serine hydroxymethyltransferase and fungal alanine racemase. A subgroup of strictly related enzymes specialized for different functions. *Eur. J. Biochem./FEBS J* **268**:6508 – 6525.

[26] Di Salvo, M.L., Florio, R., Paiardini, A., Vivoli, M., DàAguanno, S. and Contestabile, R. (2012) Alanine Racemase from Tolypocladium Inflatum: a Key PLP-dependent Enzyme in Cyclosporin Biosynthesis and a Model of Catalytic Promiscuity. *Arch. Biochem. Biophys. in press.*

[27] Kimura, T., Vassilev, V.P., Shen, G-J., Wong C-H. (1997) Enzymatic Synthesis of β-Hydroxy-α-amino Acids Based on Recombinant D- and L-Threonine Aldolases. *J. Am. Chem. Soc.* **119**(49):11734 – 11742.
doi: 10.1021/ja9720422.

[28] Karl, J., Hale, K.J., Manaviazar, S., and Delisser, V.M. (1994) A practical new asymmetric synthesis of (2S,3S)- and (2R,3R)-3-hydroxyleucine. *Tetrahedron* **50**(30):9181 – 9188.
doi: 10.1016/S0040-4020(01)85384-9.

[29] Gu, W., and Silverman, R.B. (2011) Stereospecific total syntheses of proteasome inhibitors omuralide and lactacystin. *The Journal of Organic Chemistry* **76**:8287 – 8293.
doi: 10.1021/jo201453x.

[30] Guzman-Martinez, A., Lamer, R., and Van Nieuwenhze, M.S. (2007) Total synthesis of lysobactin. *J. Am. Chem. Soc.* **129**:6017 – 6021.
doi: 10.1021/ja067648h.

[31] Morrison, R. B., and Scott, A. (1966) Swarming of Proteus – a solution to an old problem? *Nature* **211**:255 – 257.
doi: 10.1038/211255a0.

[32] Maki, H., Miura, K., and Yamano, Y. (2001) Katanosin B and plusbacin A(3), inhibitors of peptidoglycan synthesis in methicillin-resistant *Staphylococcus aureus*. *Antimicrobial agents and chemotherapy* **45**:1823 – 1827.
doi: 10.1128/AAC.45.6.1823-1827.2001.

[33] Dickinson, L., and Thompson, M.J. (1957) [The antiviral action of threo-beta-phenylserine]. *British journal of pharmacology and chemotherapy* **12**:66 – 73.
doi: 10.1111/j.1476-5381.1957.tb01364.x.

[34] van Andel, O.M. (1966) Mode of Action of L-Threo-β-phenylserine as a Chemo-therapeutant of *Cucumber Scab. Nature* **211**:326 – 327.
doi: 10.1038/211326a0.

[35] Goldstein, D.S. (2006) L-Dihydroxyphenylserine (L-DOPS): a norepinephrine pro-drug. *Cardiovascular drug reviews* **24**:189 – 203.
doi: 10.1111/j.1527-3466.2006.00189.x.

[36] Schirch, L., and Peterson, D. (1980) Purification and properties of mitochondrial serine hydroxymethyltransferase. *J. Biol. Chem.* **255**:7801 – 7806.

[37] Shapiro, A.L., Scharff, M.D., Maizel, J.V., and Uhr, J.W. (1966) Synthesis of excess light chains of gamma globulin by rabbit lymph node cells. *Nature* **211**:243 – 245.
doi: 10.1038/211243a0.

[38] Brenner, S., and Milstein, C. (1966) Origin of antibody variation. *Nature* **211**:242 – 243.
doi: 10.1038/211242a0.

[39] Sagui, F., Conti, P., Roda, G., Contestabile, R., Riva, S. (2008) Enzymatic synthesis of ω-carboxy-β-hydroxy-(L)-α-amino acids. *Tetrahedron* **64**(22):5079 – 5084.
doi: 10.1016/j.tet.2008.03.070.

[40] Fullone, M.R., Paiardini, A., Miele, R., Marsango, S., Gross, D.C., Omura, S., Ros-Herrera, E., Bonaccorsi di Patti, M.C., Lagana, A., Pascarella, S., and Grgurina, I. (2012) Insight into the structure-function relationship of the nonheme iron halo-genases involved in the biosynthesis of 4-chlorothreonine – Thr3 from *Streptomyces* sp. OH-5093 and SyrB2 from Pseudomonas syringae pv. syringae B301DR. *FEBS J.* **279**(23):4269 – 4282.
doi: 10.1111/febs.12017.

[41] Yoshida, H., Arai, N., Sugoh, M., Iwabuchi, J., Shiomi, K., Shinose, M., Tanaka, Y., and Omura, S. (1994) 4-chlorothreonine, a herbicidal antimetabolite produced by *Streptomyces* sp. OH-5093. *The Journal of Antibiotics* **47**:1165 – 1166.
doi: 10.7164/antibiotics.47.1165.

[42] Webb, H.K., and Matthews, R.G. (1995) 4-Chlorothreonine is substrate, mechanistic probe, and mechanism-based inactivator of serine hydroxymethyltransferase. *J. Biol. Chem.* **270**:17204 – 17209.
doi: 10.1074/jbc.270.29.17204.

[43] Lackner, H., Bahner, I., Shigematsu, N., Pannell, L.K., and Mauger, A.B. (2000) Structures of five components of the actinomycin Z complex from *Streptomyces fradiae*, two of which contain 4-chlorothreonine. *Journal of Natural Products* **63**:352 – 356.
doi: 10.1021/np990416u.

[44] di Salvo, M.L., Contestabile, R., and Safo, M.K. (2011) Vitamin B(6) salvage enzymes: mechanism, structure and regulation. *Biochim. Biophys. Acta* **1814**:1597 – 1608.
doi: 10.1016/j.bbapap.2010.12.006.

[45] Kotch, F.W., Guzei, I.A., and Raines, R.T. (2008) Stabilization of the collagen triple helix by O-methylation of hydroxyproline residues. *J. Am. Chem. Soc.* **130**:2952 – 2953.
doi: 10.1021/ja800225k.

[46] Wegenhart, B., Li Tan, L., Held, M., Kieliszewski, M., and Chen, L. (2006) Aggregate structure of hydroxyproline-rich glycoprotein (HRGP) and HRGP assisted dispersion of carbon nanotubes. *Nanoscale Research Letters* **1**:154 – 159.
doi: 10.1007/s11671-006-9006-8.

[47] Buczek, O., Bulaj, G., and Olivera, B.M. (2005) Conotoxins and the posttranslational modification of secreted gene products. *Cellular and Molecular Life Sciences* **62**:3067 – 3079.
doi: 10.1007/s00018-005-5283-0.

[48] Taylor, S.W., Waite, J.H., Ross, M.M., Shabanowitz, J., Hunta, D.F. (1994) trans-2,3-cis-3,4-Dihydroxyproline, a New Naturally Occurring Amino Acid, Is the Sixth Residue in the Tandemly Repeated Consensus Decapeptides of an Adhesive Protein from *Mytilus edulis*. *J. Am. Chem. Soc.* **116**:10803 – 10804.
doi: 10.1021/ja00102a063.

[49] Nakajima, T., and Volcani, B.E. (1969) 3,4-dihydroxyproline: a new amino acid in diatom cell walls. *Science* **164**:1400 – 1401.
doi: 10.1126/science.164.3886.1400.

[50] Buku, A., Faulstich, H., Wieland, T., and Dabrowski, J. (1980) 2,3-trans-3,4-trans-3,4-Dihydroxy-L-proline: An amino acid in toxic peptides of Amanita virosa mushrooms. *Proceedings of the National Academy of Sciences of the U.S.A.* **77**:2370 – 2371.
doi: 10.1073/pnas.77.5.2370.

[51] Katz, E., Mason, K.T., and Mauger, A.B. (1975) 3-hydroxy-5-methylproline, a new amino acid identified as a component of actinomycin Z1. *Biochem. Biophys. Res. Commun.* **63**:502 – 508.
doi: 10.1016/0006-291X(75)90716-0.

[52] Kaldor, S.W., Kalish, V.J., Davies, J.F., 2nd, Shetty, B.V., Fritz, J.E., Appelt, K., Burgess, J.A., Campanale, K.M., Chirgadze, N.Y., Clawson, D.K., Dressman, B.A., Hatch, S.D., Khalil, D.A., Kosa, M.B., Lubbehusen, P.P., Muesing, M.A., Patick, A.K., Reich, S.H., Su, K.S., and Tatlock, J.H. (1997) Viracept (nelfinavir mesylate,

AG1343): a potent, orally bioavailable inhibitor of HIV-1 protease. *J. Med. Chem.* **40**:3979 – 3985.
doi: 10.1021/jm9704098.

[53] Chow, W.A., Jiang, C., and Guan, M. (2009) Anti-HIV drugs for cancer therapeutics: back to the future? *The Lancet Oncology* **10**:61 – 71.
doi: 10.1016/S1470-2045(08)70334-6.

[54] Simmaco, M., Kreil, G., and Barra, D. (2009) Bombinins, antimicrobial peptides from Bombina species. *Biochim. Biophys. Acta* **1788**:1551 – 1555.
doi: 10.1016/j.bbamem.2009.01.004.

[55] Wade, D., Boman, A., Wahlin, B., Drain, C.M., Andreu, D., Boman, H.G., and Merrifield, R. B. (1990) All-D amino acid-containing channel-forming antibiotic peptides. *Proceedings of the National Academy of Sciences of the U.S.A.* **87**:4761 – 4765.
doi: 10.1073/pnas.87.12.4761.

[56] Kim, P.M., Duan, X., Huang, A.S., Liu, C.Y., Ming, G.L., Song, H., and Snyder, S.H. (2010) Aspartate racemase, generating neuronal D-aspartate, regulates adult neurogenesis. *Proceedings of the National Academy of Sciences of the U.S.A.* **107**:3175 – 3179.
doi: 10.1073/pnas.0914706107.

[57] Crow, J.P., Marecki, J.C., and Thompson, M. (2012) D-Serine Production, Degradation, and Transport in ALS: Critical Role of Methodology. *Neurology Research International,* 625245.

[58] Imai, K., Fukushima, T., Santa, T., Homma, H., Huang, Y., Sakai, K., and Kato, M. (1997) Distribution of free D-amino acids in tissues and body fluids of vertebrates. *Enantiomer* **2**:143 – 145.

[59] Misono, H., Kato, I., Packdibamrung, K., Nagata, S., and Nagasaki, S. (1993) NADP(+)-dependent D-threonine dehydrogenase from *Pseudomonas cruciviae* IFO 12047. *Applied and Environmental Microbiology* **59**:2963 – 2968.

[60] Lengsfeld, A.M., Low, I., Wieland, T., Dancker, P., and Hasselbach, W. (1974) Interaction of phalloidin with actin. *Proceedings of the National Academy of Sciences of the U.S.A.* **71**:2803 – 2807.
doi: 10.1073/pnas.71.7.2803.

[61] Liu, J. Q., Dairi, T., Itoh, N., Kataoka, M., and Shimizu, S. (2003) A novel enzyme, D-3-hydroxyaspartate aldolase from *Paracoccus denitrificans* IFO 13301: purification, characterization, and gene cloning. *Applied Microbiology and Biotechnology* **62**:53 – 60.
doi: 10.1007/s00253-003-1238-2.

[62]   Hallen, H.E., Luo, H., Scott-Craig, J.S., and Walton, J.D. (2007) Gene family encoding the major toxins of lethal Amanita mushrooms. *Proceedings of the National Academy of Sciences of the U.S.A.* **104**:19097 – 19101.
       doi: 10.1073/pnas.0707340104.

[63]   Rato, C., Amirova, S.R., Bates, D.G., Stansfield, I., and Wallace, H.M. (2011) Translational recoding as a feedback controller: systems approaches reveal polyamine-specific effects on the antizyme ribosomal frameshift. *Nucleic Acids Research* **39**:4587 – 4597.
       doi: 10.1093/nar/gkq1349.

[64]   Gijsen, H.J., Qiao, L., Fitz, W., and Wong, C.H. (1996) Recent Advances in the Chemoenzymatic Synthesis of Carbohydrates and Carbohydrate Mimetics. *Chemical Reviews* 96:443 – 474.
       doi: 10.1021/cr950031q.

[65]   Liu, J.Q., Dairi, T., Kataoka, M., Shimizu, S., and Yama, H. (2000) Diversity of microbial threonine aldolases and their application. *Mol. Catal. B. Enzym.* **10**:107 – 115.
       doi: 10.1016/S1381-1177(00)00118-1.

[66]   Duckers, N., Baer, K., Simon, S., Groger, H., and Hummel, W. (2010) Threonine aldolases-screening, properties and applications in the synthesis of non-proteinogenic beta-hydroxy-alpha-amino acids. *Applied Microbiology and Biotechnology* **88**:409 – 424.
       doi: 10.1007/s00253-010-2751-8.

[67]   Liu, J.Q., Odani, M., Dairi, T., Itoh, N., Shimizu, S., and Yamada, H. (1999) A new route to L-threo-3-[4-(methylthio)phenylserine], a key intermediate for the synthesis of antibiotics: recombinant low-specificity D-threonine aldolase-catalyzed stereospecific resolution. *Applied Microbiology and Biotechnology* **51**:586 – 591.
       doi: 10.1007/s002530051436.

[68]   Liu, J.Q., Odani, M., Yasuoka, T., Dairi, T., Itoh, N., Kataoka, M., Shimizu, S., and Yamada, H. (2000) Gene cloning and overproduction of low-specificity D-threonine aldolase from *Alcaligenes xylosoxidans* and its application for production of a key intermediate for parkinsonism drug. *Applied Microbiology and Biotechnology* **54**:44 – 51.
       doi: 10.1007/s002539900301.

[69]   Vassilev, V.P., Uchiyama, T., Kajimoto, T., and Wong, C.H. (1995) L-threonine aldolase in organic synthesis-preaparation of novel beta-hydroxy-alpha-amino acids. *Tetrahedron Lett.* **36**:4081 – 4084.
       doi: 10.1016/0040-4039(95)00720-W.

[70]   Kimura, T., Vassilev, V.P., Shen, G.J., and Wong, C.H. (1997) Enzymatic synthesis of beta-hydroxy-alpha-amino acods based on recombinant D- and L-threonine aldolases. *Journal of the American Chemical Society* **119**:11734 – 11742. doi: 10.1021/ja9720422.

[71]   Steinreiber, J., Schurmann, M., Wolberg, M., van Assema, F., Reisinger, C., Fesko, K., Mink, D., and Griengl, H. (2007) Overcoming thermodynamic and kinetic limitations of aldolase-catalyzed reactions by applying multienzymatic dynamic kinetic asymmetric transformations. *Angew. Chem. Int. Ed. Engl.* **46**:1624 – 1626.

[72]   Liu, J.Q., Ito, A., Dairi, T., Itoh, A., Shimizu, S., and Yamada, H. (1998) Low-specificity L-threonine aldolase of *Pseudomonas sp.* NCIMB 10558: purification, characterization and its application to beta-hydroxy-alpha-amino acids synthesis. *Applied Microbiology and Biotechnology* **49.**

[73]   Tanaka, T., Tsuda, C., Miura, T., Inazu, T., Tsuji, S., Nishihara, S., Hisamatsu, M., and Kajimoto, T. (2004) Design and synthesis of peptide mimetics of GDP-fucose: targeting inhibitors of fucosyltransferases. *Synlett* **2.**

[74]   Shibata, K., Shingu, K., Vassilev, V. P., Nishide, K., Fujita, T., Node, M., Kajimoto, T., and Wong, C.H. (1996) Kinetic and thermodynamic control of L-threonine aldolase catalyzed reaction and its application to the synthesis of mycestericin D. *Tetrahedron Letters* **37**:2791 – 2794. doi: 10.1016/0040-4039(96)00430-3.

[75]   Gwon, H.J., Yoshioka, H., Song, N.E., Kim, J.H., Song, Y.R., Jeong, D.Y., and Baik, S.H. (2012) Optimal production of L-threo-2,3-dihydroxyphenylserine (L-threo-DOPS) on a large scale by diastereoselectivity-enhanced variant of L-threonine aldolase expressed in *Escherichia coli. Preparative Biochemistry & Biotechnology* **42**:143 – 154. doi: 10.1080/10826068.2011.583975.

[76]   Gwon, H.J., and Baik, S.H. (2010) Diastereoselective synthesis of L-threo-3,4-dihydroxyphenylserine by low-specific L-threonine aldolase mutants. *Biotechnology Letters* **32**:143 – 149. doi: 10.1007/s10529-009-0125-z.

[77]   Giger, L., Toscano, M.D., Madeleine, B., Marlierre, P., and Hilvert, D. (2012) A novel genetic selection system for PLP-dependent threonine aldolases. *Tetrahedron* **68**:7549 – 7557. doi: 10.1016/j.tet.2012.05.097.

[78]   Steinreiber, J., Schurmann, M., van Assema, F., Wolberg, M., Fesko, K., Reisinger, C., Mink, D., and Griengl, H. (2007) Synthesis of aromatic L-2-amino alcohols using a bienzymatic dynamic kinetic asymmetric transformation. *Adv. Synth. Catal.* **349**:1379 – 1386.
doi: 10.1002/adsc.200700051.

[79]   Fesko, K., Uhl, M., Steinreiber, J., Gruber, K., and Griengl, H. (2010) Biocatalytic access to alpha,alpha-dialkyl-alpha-amino acids by a mechanism-based approach. *Angew. Chem. Int. Ed. Engl.* **49**:121 – 124.
doi: 10.1002/anie.200904395.

[80]   Fesko, K., Giger, L., and Hilvert, D. (2008) Synthesis of beta-hydroxy-alpha-amino acids with a reengineered alanine racemase. *Bioorganic & Medicinal Chemistry Letters* **18**:5987 – 5990.
doi: 10.1016/j.bmcl.2008.08.031.

[81]   Toscano, M.D., Muller, M.M., and Hilvert, D. (2007) Enhancing activity and controlling stereoselectivity in a designed PLP-dependent aldolase. *Angew. Chem. Int. Ed. Engl.* **46**:4468 – 4470.
doi: 10.1002/anie.200700710.

[82]   Seebeck, F.P., Guainazzi, A., Amoreira, C., Baldridge, K.K., and Hilvert, D. (2006) Stereoselectivity and expanded substrate scope of an engineered PLP-dependent aldolase. *Angew. Chem. Int. Ed. Engl.* **45**:6824 – 6826.
doi: 10.1002/anie.200602529.

[83]   Seebeck, F.., and Hilvert, D. (2003) Conversion of a PLP-dependent racemase into an aldolase by a single active site mutation. *Journal of the American Chemical Society* **125**:10158 – 10159.
doi: 10.1021/ja036707d.

[84]   Baik, S.H., and Yoshioka, H. (2009) Enhanced synthesis of L-threo-3,4-dihydroxy-phenylserine by high-density whole-cell biocatalyst of recombinant L-threonine aldolase from *Streptomyces avelmitilis*. *Biotechnology Letters* **31**:443 – 448.
doi: 10.1007/s10529-008-9885-0.

[85]   Nagasawa, T., Utagawa, T., Goto, J., Kim, C. J., Tani, Y., Kumagai, H., and Yamada, H. (1981) Syntheses of L-tyrosine-related amino acids by tyrosine phenol-lyase of *Citrobacter intermedius*. *Eur. J. Biochem.* **117**:33 – 40.
doi: 10.1111/j.1432-1033.1981.tb06299.x.

[86]   Phillips, R.S., Chen, H.Y., and Faleev, N.G. (2006) Aminoacrylate intermediates in the reaction of *Citrobacter freundii* tyrosine phenol-lyase. *Biochemistry* **45**:9575 – 9583.
doi: 10.1021/bi060561o.

[87] Koyanagi, T., Katayama, T., Suzuki, H., Onishi, A., Yokozeki, K., and Kumagai, H. (2009) Hyperproduction of 3,4-dihydroxyphenyl-L-alanine (L-Dopa) using *Erwinia herbicola* cells carrying a mutant transcriptional regulator TyrR. *Bioscience, Biotechnology, and Biochemistry* **73**:1221 – 1223.
doi: 10.1271/bbb.90019.

[88] Phillips, R.S., Demidkina, T.V., and Faleev, N.G. (2003) Structure and mechanism of tryptophan indole-lyase and tyrosine phenol-lyase. *Biochim. Biophys. Acta* **1647**:167 – 172.
doi: 10.1016/S1570-9639(03)00089-X.

[89] Miles, E.W. (2013) The tryptophan synthase alpha2beta2 complex: a model for substrate channeling, allosteric communication, and pyridoxal phosphate catalysis. *J. Biol. Chem.* **288**:10084 – 10091.
doi: 10.1074/jbc.X113.463331.

[90] Phillips, R.S., Cohen, L.A., Annby, U., Wensbo, D., Gronowitz, S. (1995) Enzymatic synthesis of Thia-L-tryptophans. *Bioorganic & Medicinal Chemistry Letters* **5**(11):1103 – 1206.
doi: 10.1016/0960-894X(95)00181-R.

[91] Milton, J.S., Phillips, R.S. (1992) Enzymatic synthesis of aza-l-tryptophans: The preparation of 5- and 6-Aza-L-tryptophan. *Bioorganic & Medicinal Chemistry Letters* **2**:1053 – 1056.
doi: 10.1016/S0960-894X(00)80617-4.

[92] Lee, M. *et al.* (1992) Enzymatic synthesis of chloro-L-tryptophans. *Bioorganic & Medicinal Chemistry Letters* **2**:1563 – 1564.

[93] Rabeh, W.M., and Cook, P.F. (2004) Structure and mechanism of O-acetylserine sulfhydrylase. *J. Biol. Chem.* **279**:26803 – 26806.
doi: 10.1074/jbc.R400001200.

[94] Mozzarelli, A., Bettati, S., Campanini, B., Salsi, E., Raboni, S., Singh, R., Spyrakis, F., Kumar, V.P., and Cook, P.F. (2011) The multifaceted pyridoxal 5'-phosphate-dependent O-acetylserine sulfhydrylase. *Biochim. Biophys. Acta* **1814**:1497 – 1510.
doi: 10.1016/j.bbapap.2011.04.011.

[95] Maier, T.H. (2003) Semisynthetic production of unnatural L-alpha-amino acids by metabolic engineering of the cysteine-biosynthetic pathway. *Nature Biotechnology* **21**:422 – 427.
doi: 10.1038/nbt807.

[96]   Omura, H., Kuroda, M., Kobayashi, M., Shimizu, S., Yoshida, T., and Nagasawa, T. (2003) Purification, characterization and gene cloning of thermostable O-acetyl-L-serine sulfhydrylase forming beta-cyano-L-alanine. *Journal of Bioscience and Bioengineering* **95**:470 – 475.

[97]   Yamagata, S. (1989) Roles of O-acetyl-L-homoserine sulfhydrylases in micro-organisms. *Biochimie* **71**:1125 – 1143.
        doi: 10.1016/0300-9084(89)90016-3.

[98]   Chocat, P., Esaki, N., Tanaka, H. and Soda, K. (1985) Synthesis of Selenocystine and Selenohomocystine with O-Acetylhomoserine Sulfhydrylase. *Agric. Biol. Chem.* **49**(4):1143 – 1150.
        doi: 10.1271/bbb1961.49.1143.

[99]   Omura, H., Ikemoto, M., Kobayashi, M., Shimizu, S., Yoshida, T., and Nagasawa, T. (2003) Purification, characterization and gene cloning of thermostable O-acetyl-L-homoserine sulfhydrylase forming gamma-cyano-alpha-aminobutyric acid. *Journal of Bioscience and Bioengineering* **96**:53 – 58.

[100]   Tran, T.H., Krishnamoorthy, K., Begley, T.P., and Ealick, S.E. (2011) A novel mechanism of sulfur transfer catalyzed by O-acetylhomoserine sulfhydrylase in the methionine-biosynthetic pathway of *Wolinella succinogenes*. *Acta Crystallogr. D Biol. Crystallogr.* **67**:831 – 838.
        doi: 10.1107/S0907444911028010.

[101]   Budisa, N. (2004) Prolegomena to future experimental efforts on genetic code engineering by expanding its amino acid repertoire. *Angew. Chem. Int. Ed. Engl.* **43**:6426 – 6463.
        doi: 10.1002/anie.200300646.

[102]   Kobayashi, T., Yanagisawa, T., Sakamoto, K., and Yokoyama, S. (2009) Recognition of non-alpha-amino substrates by pyrrolysyl-tRNA synthetase. *Journal of Molecular Biology* **385**:1352 – 1360.
        doi: 10.1016/j.jmb.2008.11.059.

[103]   Budisa, N., Wenger, W., and Wiltschi, B. (2010) Residue-specific global fluorination of Candida antarctica lipase B in *Pichia pastoris*. *Molecular bioSystems* **6**:1630 – 1639.
        doi: 10.1039/c002256j.

[104]   Young, T.S., and Schultz, P.G. (2010) Beyond the canonical 20 amino acids: expanding the genetic lexicon. *J. Biol. Chem.* **285**:11039 – 11044.
        doi: 10.1074/jbc.R109.091306.

[105] Budisa, N., Minks, C., Alefelder, S., Wenger, W., Dong, F., Moroder, L., and Huber, R. (1999) Toward the experimental codon reassignment *in vivo*: protein building with an expanded amino acid repertoire. *FASEB J.* **13**:41–51.

[106] Furter, R. (1998) Expansion of the genetic code: site-directed p-fluoro-phenylalanine incorporation in *Escherichia coli*. *Protein Science* **7**:419–426. doi: 10.1002/pro.5560070223.

[107] Wang, L., and Schultz, P. G. (2001) A general approach for the generation of orthogonal tRNAs. *Chemistry & Biology* **8**:883–890. doi: 10.1016/S1074-5521(01)00063-1.

[108] Angelaccio, S. (2013) Extremophilic SHMTs: from structure to biotecnology. *Bio.Med. Research International, in press.*

[109] Welch, M., and Phillips, R. S. (1999) Enzymatic syntheses of 6-(4 H-selenolo[3,2-b] pyrrolyl)-L-alanine, 4-(6 H-selenolo[2,3-b]pyrrolyl)-L-alanine, and 6-(4 H-furo[3,2-b] pyrrolyl-L-alanine. *Bioorg. Med. Chem. Lett.* **9**:637–640. doi: 10.1016/S0960-894X(99)00067-0.

[110] Lepthien, S., Hoesl, M.G., Merkel, L., and Budisa, N. (2008) Azatryptophans endow proteins with intrinsic blue fluorescence. *Proceedings of the National Academy of Sciences of the U.S.A.* **105**:16095–16100. doi: 10.1073/pnas.0802804105.

[111] Budisa, N. (2006) *Engineering the Genetic Code: Expanding the Amino Acid Repertoire for the Design of Novel Proteins*, Wiley-VCH Verlag GmbH, Weinheim.

Messages from Nature – How CHEMICAL Synthetic Biology Could Look Like

Andreas Kirschning*, Olena Mancuso, Jekaterina Hermane, and Gerrit Jürjens

Institute of Organic Chemistry
and Center of Biomolecular Drug Research (BMWZ),
Leibniz University Hannover, Schneiderberg 1b, 30167 Hannover, Germany

E-Mail: *andreas.kirschning@oci.uni-hannover.de

Received: 7th January 2013 / Published: 13th December 2013

Introduction

The term synthetic biology is a relative new expression for an emerging topic at the interface between chemistry and biology. Other, related fields may include medicinal chemistry, bioorganic chemistry and chemical biology; all these subfields are still based on chemical synthesis. Likewise one can see this trend of diversification in the field of biotechnology which has been complemented with terms like metabolic engineering or combinatorial biosynthesis. Many of these terms are not that strictly defined and substantially overlap and redundancy has to be acknowledged. There is still some debate of what synthetic biology covers. In short, one may describe it to assemble and merge genes *in vitro* to create a new useful biological system [1, 2].

The field of synthetic biology is mainly driven by biologists and biotechnologists. It seems as if in these early days of this modern type of biotechnology the expertise of synthetic chemists as well as the opportunities associated with chemical synthesis are overlooked or ignored. The question may be raised whether the *de novo* establishment of biosynthetic pathways will consequently lead to new derivatives of a target molecule or will only provide a better access to a known natural product. One can also doubt that by including enzymatic cassettes that are not on the given natural biosynthetic pathway will provide a wide variety of new derivatives. If derivatives are required e. g. for improving certain desirable properties particularly in the arena of medicinal chemistry, chemical synthesis still provides the largest flexibility.

In scheme 1 this is exemplified for artemisinin, a diterpene that has emerged as one of the most effective antimalarial drugs [3]. It is produced by *Artemisia annua* but isolated in amounts that makes this drug rather costly in view of the economic situation in countries where malaria is most abundant [4]. Keasling *et al.* reported on the production of the antimalarial drug precursor artemisinic acid in engineered yeast which has become one of the representative examples for synthetic biology and can be regarded to be one of the initial milestones [5]. However, their work provided a biosynthetic precursor, namely artemisinic acid which is not active as an antimalarial drug. It was the Seeberger group [6] that recently provided an elegant flow approach relying on a short chemical sequence that likely will pave the way to produce artemisinin using Keasling's engineered organism and Seeberger's semisynthesis. This example illustrates without adressing the issue of accessing new derivatives, how synthetic biology or metabolic engineering will beneficially work hand in hand with chemical synthesis to practically solve synthetic challenges like the medically important natural product artemisinin.

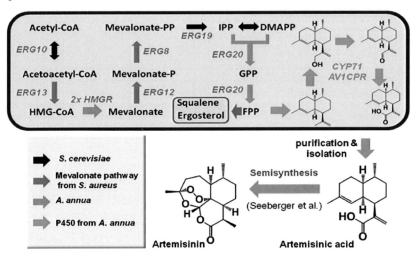

Scheme 1. Production of artemisinin by combining of biosynthetic modification of *Saccheromyces cerevisiae* with semisynthesis.

Indeed, natural products have recently proven to open up new avenues in emerging areas of biomedical research. Gupta and coworkers studied the effect of small molecules on the development of cancer stem cells [7]. These cells have become a hot topic in oncological research because their resistance towards current cancer therapies including chemo- and radiation therapy [8]. For that purpose a compound library containing about 16.000 entities was screened for its selective inhibitory properties on cancer stem cells. The assay revealed

32 active compounds from which four natural products, namely Salinomycin (1), Abamectin (2), Nigericin (3) and Etoposide (4) (Fig. 1) turned out to be the most active ones. Noteworthy, except for Etoposide, all of these known molecules are polyketide-derived.

In this report selected showcases from our laboratories which are far from covering the broad scope of possible and thinkable strategies, will demonstrate, that the liaison of chemical synthesis and modern metabolic engineering provides great future prospects for expanding the opportunities for natural products and derivatives particularly in the field of medicinal chemistry and pharmaceutical research.

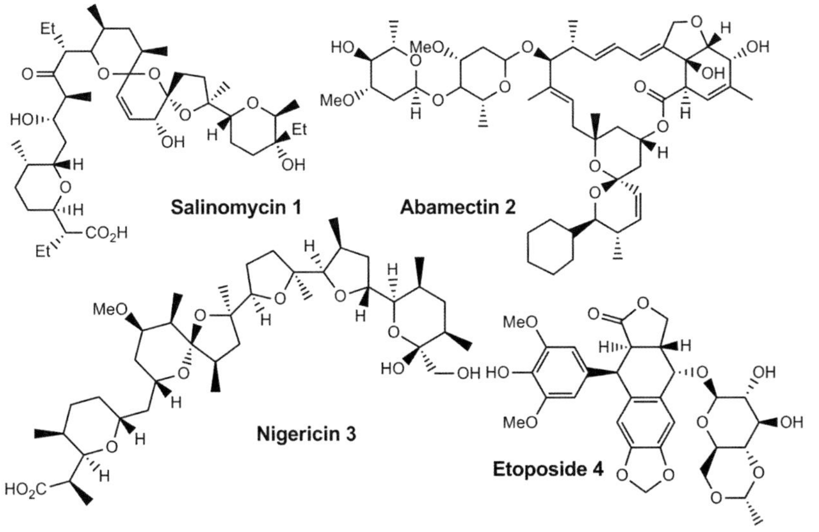

Figure 1. Structures of Salinomycin (1), Abamectin (2), Nigericin (3) and Etoposide (4).

MUTATIONAL BIOSYNTHESIS (MUTASYNTHESIS)

In 1969, Rinehart and Gottlieb introduced the term mutational biosynthesis in a rather visionary account in which they speculated on opportunities of how to manipulate biosynthetic machineries on a genetic level [9]. Their proposals and ideas were visionary because at that time the techniques of molecular biology and metabolic engineering did not exist to interfere with the genome with the precision of today's routines. In fact, these opportunities were supposed to come decades later. Mutational biosynthesis or in short mutasynthesis comprises the generation of mutants of a producer organism blocked in the formation of a key biosynthetic building block required for assembly of the end-product [10]. The administration of potentially altered building blocks, so-called mutasynthons, to the blocked mutant may result in new metabolites which are isolated and biologically evaluated (Fig. 2).

Importantly, the authors stressed the need of isolation and biological evaluation of the new derivatives formed. This is a very important aspect if mutasynthesis wants to become an established and respected method because it competes with pure chemical methods such as semi and total synthesis and only to a minor extent it also plays a role in elucidating details on biosynthetic pathways.

Nowadays, a much deeper understanding of biosynthetic pathways of complex secondary metabolites has been gained. More importantly, tools in molecular biology and genetic engineering are now in hand [11], that allow to carry out natural product synthesis similar to total synthesis programs that are flexibilized by merging chemical synthesis with bio-synthesis, so that natural products may return back on the agenda of drug development [12]. Conceptually the best of both worlds, the flexibility of chemical synthesis for introducing structural changes and the preciseness of biosynthetic pathways are chosen and combined for a given target molecule.

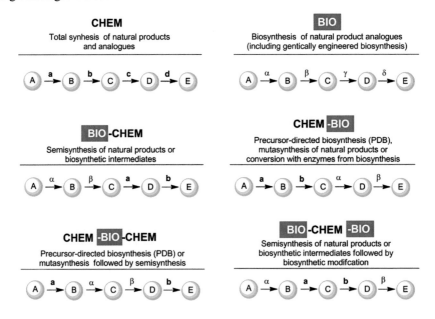

Figure 2. Classification of "total synthesis" approaches towards natural product analogues and libraries based on chemical and biological methods and combinations (these are selected combinations but other combinations like BIO-BIO-CHEM can also be envisaged; (a-d = chemical reagents or catalysts; α-δ = enzymes; A = starting material; B-D = synthetic and/or biosynthetic intermediates; E = natural product or derivative) [13].

Recently, we suggested a simple classification for visualizing the synthetic opportunities of hybrid techniques (Fig. 2) [13]. It mainly focuses on synthetic strategies. Two abbreviations serve to distinguish between chemical synthesis labelled CHEM and biotransformations performed by overexpressed enzymes or biosynthetic multienzyme cassettes called BIO.

This for example means that a BIO-CHEM synthesis is simply a semisynthetic derivatization of a starting material obtained from a natural source or by fermentation. This creates structural diversity which is further expanded by the extended BIO-CHEM-BIO approach. This would mean that a semisynthetically modified natural product is subjected to an enzymatic biotransformation or to a whole cell fermentation [13].

THE ANSAMYCIN ANTIBIOTICS

A showcase for hybrid CHEM-BIO total syntheses

An important group of secondary metabolites are the ansamycins which are macrolactams, that contain an aryl, naphtyl or quinone chromophor which is part of a polyketide-type ansa chain. The chain is attached via the meta positions to the chromophor. Famous examples are rifamycin, geldanamycin (9) and maytansin [14]. One member of an aryl ansamycin antibiotic are the maytansinoids which exhibit cytotoxic activity, evident in the growth inhibition of different tumor cell lines and human solid tumors at very low concentrations (10^{-3} to 10^7 µg/ml) [14]. Their antimitotic mode of action is based on the interaction with β-tubulin. It prevents polymerisation of tubulin and thereby promoting depolymerisation of microtubules. The maytansinoids currently attract high clinical interest as "warheads" in tumor-targeted immunoconjugates [15]. Maytansin was isolated from plant material while later the closely related ansamitocins (8), were found in the bacteria like *Actinosynnema pretiosum*. The biosynthesis of the ansamitocins involves several unusual steps of which the aromatic biosynthetic starter 3-amino-5-hydroxybenzoic acid (AHBA) (5) supplied by a dedicated biosynthetic pathway [16] (Scheme 2) is an important feature. Furthermore, a separate amide synthase acting after polyketide synthase processing provides proansamitocin (6). This cyclization is another key enzymatic process of the ansamycin antibiotics before the resulting core structure 6 is further decorated by a set of tailoring enzymes.

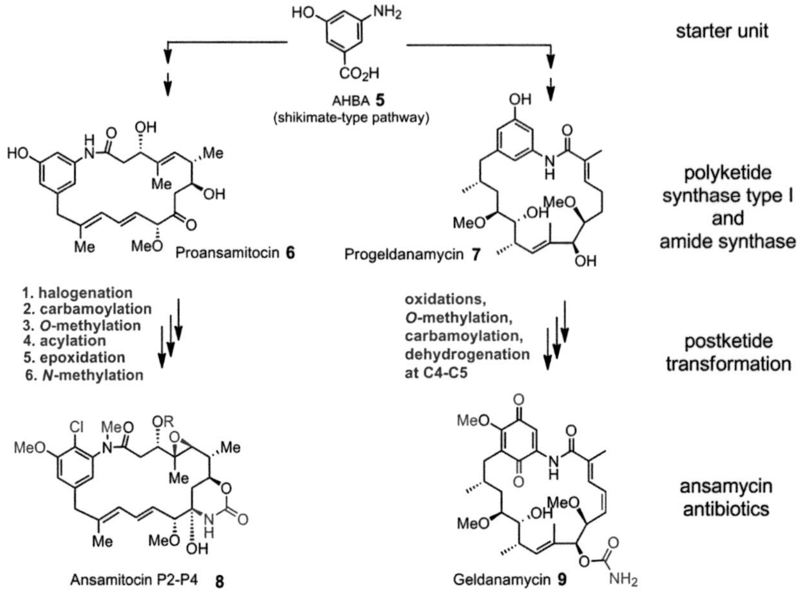

Scheme 2. Short overview over the biosynthesis of the two ansamycin antibiotics ansamitocin and geldanamycin.

Likewise, geldanamycin (**9**) is another important member of the ansamycins, and it is produced by *Streptomyces hygroscopicus* var. *geldanus*. Geldanamycin is a potential anti-tumor drug [17] which binds to the *N*-terminal ATP-binding domain of heat shock protein 90 (Hsp90) and consequently it inhibits the ATP-dependent chaperone activities [18]. Most unnatural geldanamycin derivatives reported to date are aminated in the 17 position by a semisynthetic step [19]. The biosynthesis of geldanamycin (**9**) is closely related to ansami-tocin P-3 (**8**). It is also based on a PK synthase with AHBA (**5**) as the starter unit and a stand-alone amide synthase for cyclisation. Individual post-PKS tailoring enzymes follow formation of progeldanamycin (**7**) that differ from those occuring in *A. pretiosum* terminate geldanamycin biosynthesis (Scheme 2).

Geldanamycin is the best studied example of ansamycin antibiotics being modified by mutasynthesis as well as other metabolic engineering approaches [20 – 22], while so far only our group has been involved in this arena to create libraries of ansamycin antibiotics [23, 24].

Indeed, ansamycin antibiotics are ideally suited for mutasynthetic generation of analogues as AHBA (**8**), the starter unit exerts biosynthetic uniqueness. For this purpose a knock-out strain of *A. pretiosum* has to be generated that is blocked in the biosynthesis of AHBA. By doing so, one does not interfere with the essential primary metabolic pathways in the microorganism. Gratifyingly, the AHBA loading domain of the ansamitocin PKS turned out to accept and process a structurally diverse number of 3-aminobenzoic acids as exemplified for aminobenzoic acid (10) which was smoothly transformed into ansamitocin derivative 11 (Scheme 3) [25, 26].

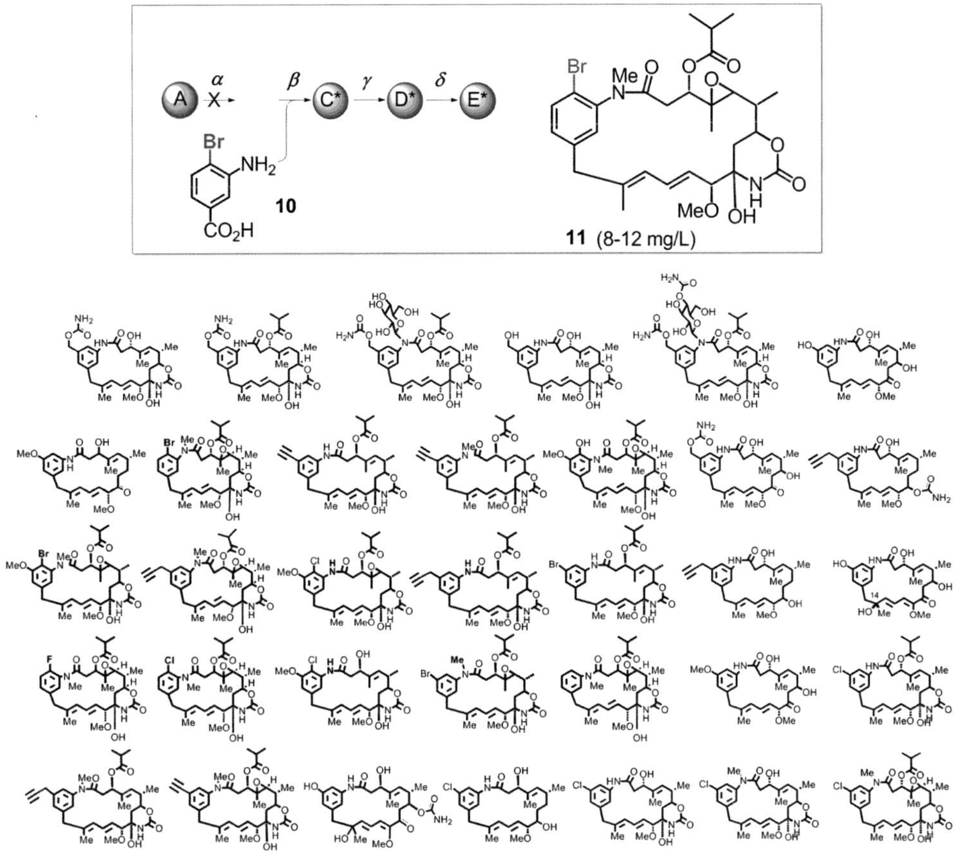

Scheme 3. Mutasynthetic preparation of ansamitocin derivative 11 from mutasynthon 10 and representative examples of new ansamiton derivatives obtained by mutasynthesis using a AHBA blocked mutant of *A. pretiosum*.

In a similar manner, the geldanamycin producer *S. hygroscopicus* var. *geldanus* was genetically manipulated and exploited in mutasynthetic fermentations by supplementing 3-aminobenzoic acid derivatives such as 12 (Scheme 4) [23]. In fact, aza-geldanamycin derivative (13) represents the first example of a mutaproduct that was generated from feeding an aromatic heterocycle to an AHBA blocked mutant.

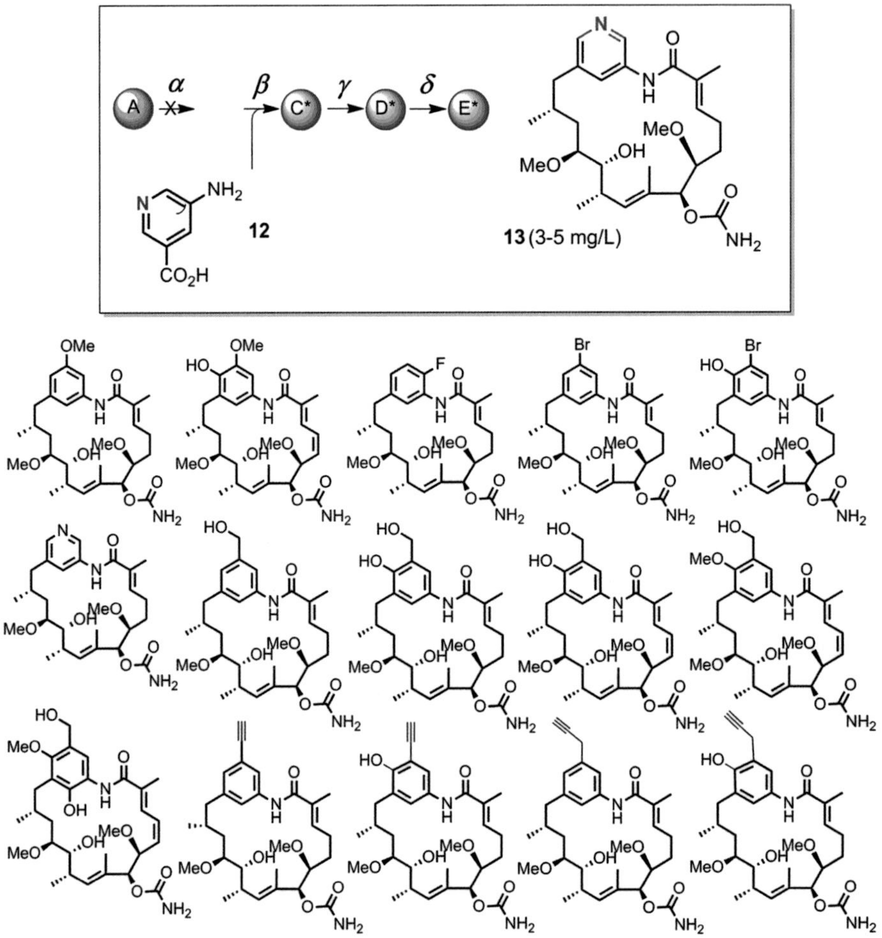

Scheme 4. Mutasynthetic preparation of aza geldanamycin derivative **13** from muta-synthon **12** and representative examples of new geldanamycin derivatives obtained by mutasynthesis using a AHBA blocked mutant of *S. hygroscopicus*.

Mutasynthesis followed by semisynthesis (CHEM-BIO-CHEM)

In order to broaden the options for diversity oriented synthesis the concept of mutasynthesis can be extended in a combination with semisynthesis (Fig. 2). This strategy requires, that the product formed from mutasynthesis bears a new, chemically useful functionality, which then serves for semisynthetic derivatizations. The CHEM-BIO-CHEM approach requires that the mutasynthesis provides sufficient amounts of the new natural product for chemical synthesis. It needs to be noted that this requirement is not always fulfilled with mutasynthesis.

An early example is shown in Scheme 5. Feeding of 3-amino-4-bromo-5-hydroxy benzoic acid **14** to the AHBA(-) blocked mutant of *A. pretiosum* [27] furnished the bromo-AP-3 derivative **15** in very good yield. Alternatively, the substrate selectivity of the perhalogenase can be extended to bromination when the fermentation broth was supplemented with sodium bromide and AHBA **5**. Bromo-AP-3 **15** is a key precursor for semisynthetic derivatisations particularly when Pd-catalysed reactions such as the Stille and the Sonogashira cross coupling reactions are exploited. With the Stille coupling the vinyl group was introduced to furnish vinyl AP-3 **16** [25]. Alkinylation of the aromatic moiety was achieved with TMS acetylene which gave ansamitocin P-3 derivative **17** (Scheme 5). It is noteworthy, that despite the substantial alterations in the aromatic unit of AP-3, two of these new ansamitocin derivatives **15** and **16** show strong cytotoxic activity (< 20 nM) towards a classical panel of cell lines.

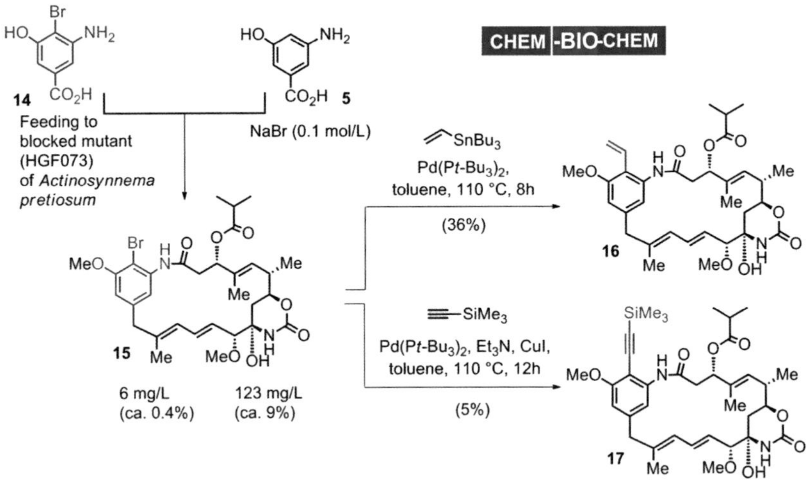

Scheme 5. Combined mutasynthesis/biosynthesis and semisynthesis for the preparation of ansamitcon derivatives **16** and **17** [25].

Proansamitocin – chemical and biochemical synthesis and feeding studies

Proansamitocin **6**, the first macrolactam biosynthetic intermediate, is another versatile starting point for performing semisynthetic derivatisations (Scheme 2). In contrast to the mutasynthetic strategies described in schemes 3 and 4 where the aromatic moiety in ansamitocin was chemically modified, proansamitocin paves the way to carry out semisynthesis on functional groups that are located in the ansa chain. In order to assure that complex mutasynthons like proansamitocin **6** and semisynthetic products derived therefrom are able to penetrate the bacterial membrane and are further processed inside, we prepared proansamitocin **6** by total synthesis and fed it to the AHBA blocked mutant of *A. pretiosum*.

The total synthesis is briefly depicted in Scheme 6. It starts from benzyl bromide **18** [28] which was converted by a two-step process into vinyl iodide **20**, involving an alkynylation using alkynylindium **19** followed by a carbomethylation and *ipso*-iodination. The diene required for the alkene-diene ring closing metathesis was set up by a Stille reaction. Next, the major part of the ketide-based ansachain **22** [29], which had been prepared by conventional stereocontrolled reactions common in polyketide synthesis, was introduced via the amino group. The resulting amide **23** was then subjected to ring closing metathesis conditions using the Grubbs 1 catalyst which after desilylation yielded proansamitocin **6** (starting material and a small amount of *E,Z* diene were also isolated). It needs to be mentioned, that this is a rare example of a complex alkene diene ring closing metathesis reaction. The key experiment, feeding of pronsamitocin **6** to the AHBA blocked mutant of *A. pretiosum* was successful in that the last steps of the biosynthesis took place and AP-3 **8b** and its dechloro derivative were formed and isolated [29]. This experiment paved the way to prepare analogues of proansamitocin in order to utilize the tailoring enzymes of ansamitocin biosynthesis.

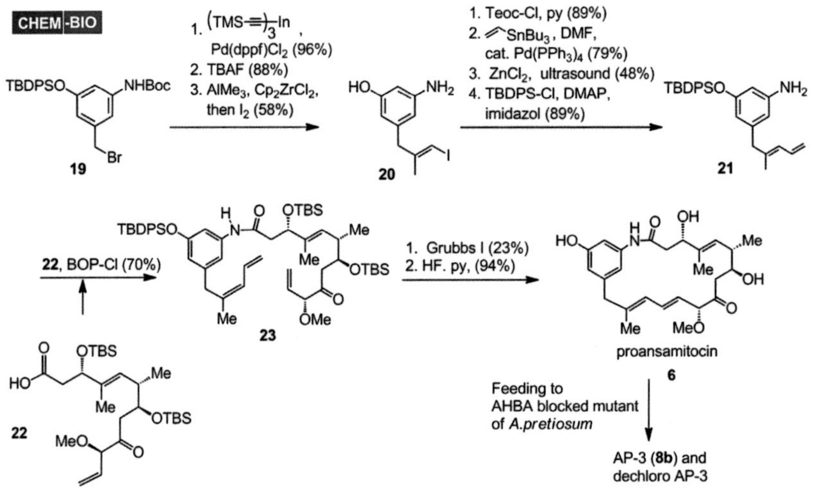

Scheme 6. Total synthesis of proansamitocin **6** and feeding to ΔAHBA blocked mutant of *A. pretiosum*.

Next, we utilised a new blocked mutant of *A. pretiosum* that is unable to carry out postketide modifications [blocked in *asm12* (chlorination) and *asm21* (carbamoylation)] which provided proansamitocin **6** [27, 30] in good yield (Fig. 3). Additionally, small amounts of 10-*epi*-proansamitocin **24** and *O*-methyl proansamitocin **25** were isolated. Very likely **24** may arise from α-epimerization of the keto group at C-9 during work-up. In addition, the *O*-methyl transferase (Asm7), that remained intact in the blocked mutant, is still able to operate on proansamitocin to a minor degree. Finally, two separable diastereomeric byproducts **26a,b** were isolated that contained an additional hydroxyl group at C-14 and the diene moiety, which may have resulted from an intermediate oxirane at C 13-C 14.

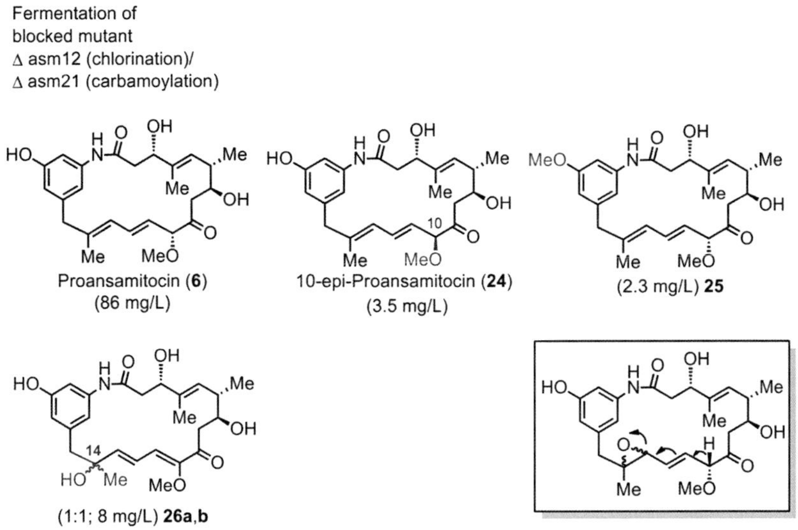

Fermentation of
blocked mutant
Δ asm12 (chlorination)/
Δ asm21 (carbamoylation)

Figure 3. Structures of proansamitocin 6, 10-*epi*-proansamitocin 24, *O*-methyl proansamitocin 25 and oxidation by-products 26a,b.

Biosynthesis followed by semisynthesis and mutasynthesis (BIO-CHEM-BIO)

With sufficient amounts of proansamitocin **6** in hand, semisynthetic derivatisation in the ansa chain of ansamitocin was pursued. For example, nucleophilic methylation of the keto group at C-9 in per-*O*-silylated proansamitocin **27** yielded the methyl-branched proansamitocins **28a,b** (5:1 for MeLi or 10:1 for MeMgBr, respectively) (Scheme 7) [27].

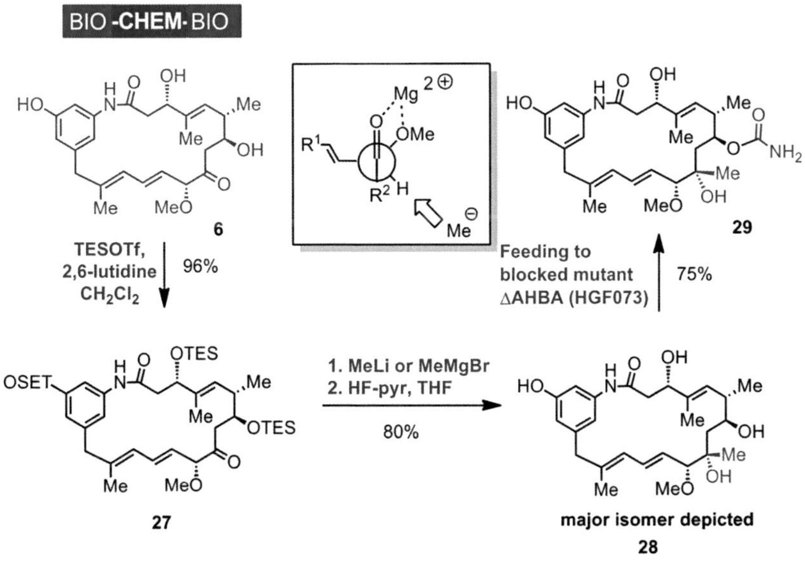

Scheme 7. Semisynthesis of C-9 branched proansamitocin derivative **29** starting from proansamitocin following a BIO-CHEM-BIO approach.

We postulate that the above conformer (framed) is enforced by the chelating properties of the neighbouring methoxy group. This chelation is expected to control the facial selectivity of the nucleophilic attack. When this derivative was supplemented to the AHBA blocked mutant of *A. pretiosum*, carbamoylation of the hydroxyl group at C7 took place and provided the proansamitocun derivative **29** in good yield. In this example two blocked mutants served as synthetic tool while flexible derivatisation was achieved by chemical synthesis.

It is well accepted that the isobutyrate group at C3 in AP-3 is essential for its strong antiproliferative activity, while it has been debated, whether the cyclic carbamate has pharmacophoric properties. Therefore, we first devised a synthetic sequence that allows to introduce esters at C-3 in proansamitocin and its analogues. Being generally applicable this two-step protocol is represented for the preparation of ester **29** starting from the major diastereomer **28a,b**. This synthetic route was exploited to address one SAR issue that had remained to be part of an open debate for several decades.

What is the pharmacological role of the cyclic carbamate moiety in AP-3?

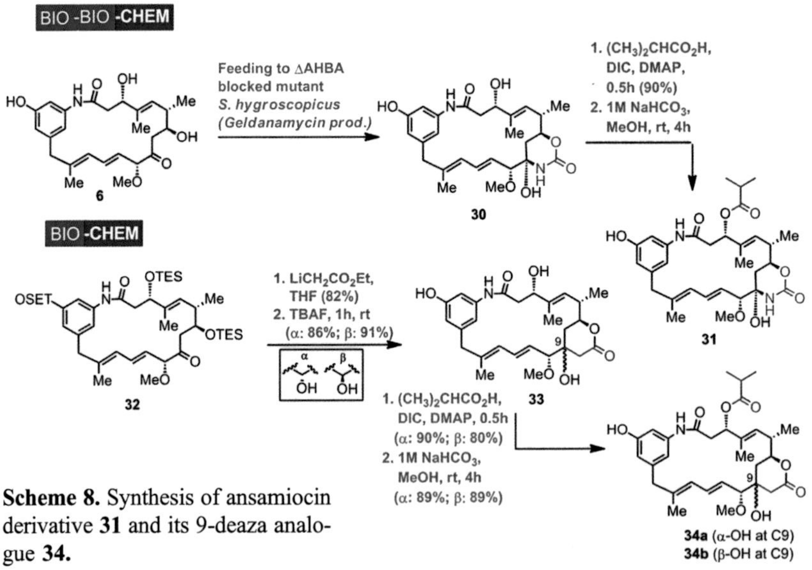

Scheme 8. Synthesis of ansamiocin derivative **31** and its 9-deaza analogue **34**.

Thus, besides the simplest AP-3 derivative that bears both the ester side chain as well as the carbamate moiety, we also prepared the carbaanalogues **34** by semisynthetic modification of per-*O*-silylated proansamitocin **32** (Scheme 8). First, the Claisen-reaction gave diastereomeric lactons **33a,b** after fluoride-induced desilylation. Then, acylation at C3 was carried out as developed for the preparation of carbamate **31**. As a result, this short sequence yielded the deaza analogues **34a,b**. Comparative evaluation of the antiproliferative activities of these three derivatives **31** and **34a,b** unravelled that the carbaanalogue **34a** with α-orientation of the hydroxyl group at C9 shows antiproliferative activity in the lower nanomolar range which is in the similar range as was found for the carbamate **31**. In contrast, the 9-epimer **34b** is substantially less active

Targeting tumors: Conjugate assembly by a CHEMBIOCHEM approach

The antitumor activity of the maytansinoids was extensively evaluated in human clinical trials [31], but although potent *in vitro*, the maytansinoids displayed a poor therapeutic window *in vivo*. By forming conjugates with monoclonal antibodies [32], vitamins such as folic acid [33] and others [34] that bind to specific markers on the surface of tumor cells, ansamitocins and related highly cytotoxic agents show improved selectivity towards tumors associated with reduced cytotoxicity. Recently, the humanized anti-HER2 mAb antibody trastuzumab was conjugated with maytansine and this construct reached late clinical trials for the treatment of HER2+ metastatic breast cancer [35].

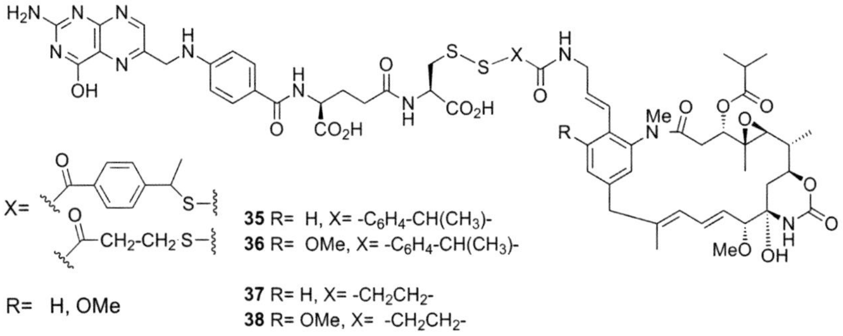

Figure 4. Folic acid/AP-3 conjugates **35-38** prepared by a combined synthetic/muta-synthetic approach.

The vitamin folic acid is a promising ligand for selective delivery of attached therapeutic agents to tumor tissues [36]. Many different cancer cells do overexpress folic acid receptors (FR) on their cell surfaces [37]. Folic acid-drug conjugates targeting FR are commonly constructed of three components, the the folic acid, the cytotoxic drug, and the linker connecting the drug to the tumor-specific ligand. The latter should be able to undergo a release mechanism of the active drug [38]. It is established that a folate-disulfide-drug conjugate can undergo reduction after endocytosis [36].

We prepared four new tumor specific folic acid/ansamitocin conjugates **35-38** (Fig. 4) and utilised a synthetic strategy based on the combination of mutasynthesis and semisynthesis. The two bromo-ansamitocin derivatives **15** and **11** are accessible by mutasynthesis or by a modified fermentation protocol, respectively (Scheme 9). An allyl amine linker **40** was introduced under Stille conditions [39] to yield modified ansamitocins **41** and **42**.

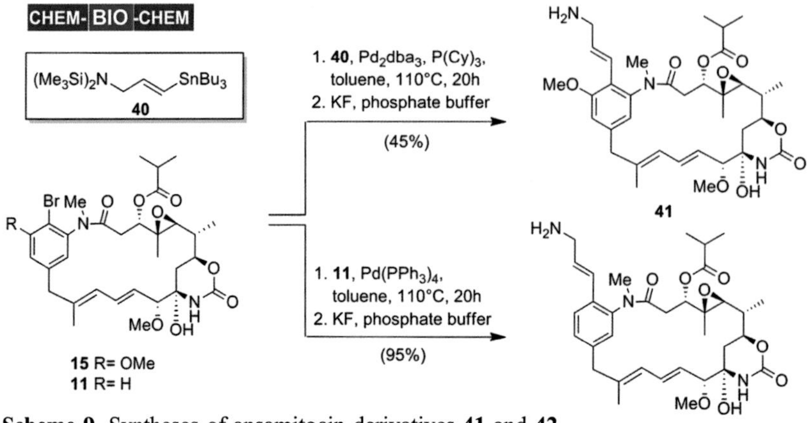

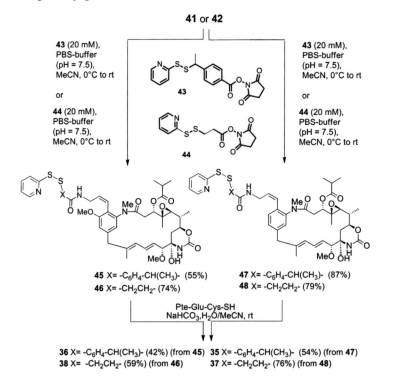

Scheme 9. Syntheses of ansamitocin derivatives **41** and **42**.

The amino group was transformed into the pyridyl disulfide derivatives **45-48** with doubly activated 3-(2-pyridyldithio)-propionic ester (SPDP) **43** and 4-[1-(2-pyridyldithio)-ethyl]-benzoic ester (SMPT) **44**, respectively (Scheme 10). Standardised coupling steps introduced the pteroic acid/glutamic acid/cysteine (Pte-Glu-Cys-SH) unit to the modified ansamitocins yielding the target conjugates **35-38**.

Scheme 10. Preparation of folic acid/AP3 conjugates **35-38**.

These conjugates as well as those that are expected to be generated after internalization of the folic acid/ansamitocin conjugates into the cancer cell and reductive cleavage of the disulfide linkage 49 – 51 [40] were evaluated with respect to their antiproliferative activity. The latter were prepared from the corresponding pyridyl disulfides after treatment with dithiothreitol (DTT) in PBS-buffered acetonitrile. All three thiols exerted good to strong antiproliferative activities against selected cell lines including a lung carcinoma A-549 devoid of membrane-bound folic acid receptors (FR) (Table 1). These modified ansamitocins are about 10 – 100 fold less active than ansamitocin P3 (8b) but are still suited to serve as an anticancer agents.

Table 1. Antiproliferative activity IC$_{50}$ [nM] of ansamitocin derivatives **49 – 51** in SW-480, A-549 and HCT-116 cells (FR-: devoid of membrane-bound folic acid receptors).

Cell line	origin	49	50	51
SW-480	colon carcinoma	5	6	42
KB-3 – 1	cervix carcinoma	4	4	14
A-549 (FR-)	lung carcinoma	6	24	420
HCT-116	colon carcinoma	8	10	49

49 R= H, X= -C$_6$H$_4$-CH(CH$_3$)-
50 R= OMe, X= -C$_6$H$_4$-CH(CH$_3$)-
51 R= H, X= -CH$_2$CH$_2$-

Conjugates **35-38** were tested to two cancer cell lines. Cervix carcinoma KB-3 – 1 (FR+) contained membrane-bound folic acid receptors whereas lung carcinoma A-459 (FR-) was deficient in this respect. All four conjugates showed strong antiroliferative activity for the former cell line but were inactive towards the latter FR- cell line (Table 2). The values of cytotoxicity favorably match with those for thiols **49-51** (Table 1) the supposed products resulting from *in vivo* reductive cleavage of **35-38**.

Table 2. Antiproliferative activity IC$_{50}$ [nM] of folate-AP-3 conjugates **22-25**. (KB-3 – 1 cells: folate receptor positive and A-549 cells: folate receptor negative).

Cell line	Origin	35	36	37	38
KB-3 – 1 (FR+)	cervix carcinoma	13	8	21	7
A-549 (FR-)	lung carcinoma	$> 10^4$	$> 10^4$	$> 10^4$	$> 10^4$

We regard this project as a proof for the power of combining mutasynthesis and semisynthesis in order to access complex natural product architectures and a proof that these conjugates are able to exert selectivity.

New macrocyclic architectures by mutaysynthesis

The combination of chemical synthesis and biotransformations using engineered organisms can even pave new avenues towards novel macrocyclic backbones. When an AHBA(-) mutant of *Streptomyces hygroscopicus var. geldanus* was supplemented with hydroxymethylaminobenzoic acid (52) several new metabolites were formed, isolated and fully characterised (Fig. 5) [24]. Five metabolites 53 – 57 can be put in a logical biogenetic order. Thus, the timing of biosynthetic events of the tailoring biotransformations can be postulated in that after ring-closure of the progeldanamycin-analogue by the amide synthase, carbamoylation, oxidation at C17, 4,5-desaturation and O-methylation take place. The final tailoring modification is the oxidation at C21. Importantly, also three additional mutaproducts 58a-58c were isolated which are 20-membered macrolactones, compared to the 19-membered macrolactam ring in geldanamycin (9) (Fig. 6). Clearly, the amide synthase, that is responsible for macrolactam formation, is also able to cyclise the PKS-bound seco acid via the mere nucleophilic benzylic alcohol.

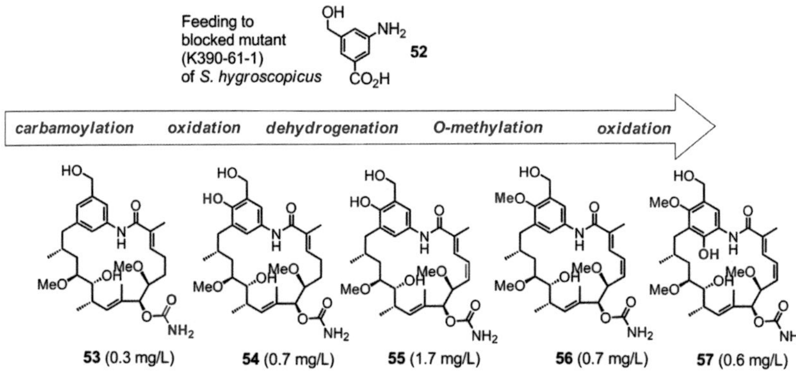

Figure 5. Mutasynthesis with *S. hygroscopicus* (strain K390 – 61 – 1) using 3-amino-5-hydroxymethylbenzoic acid 52 and the proposed sequence of tailoring biotransformations.

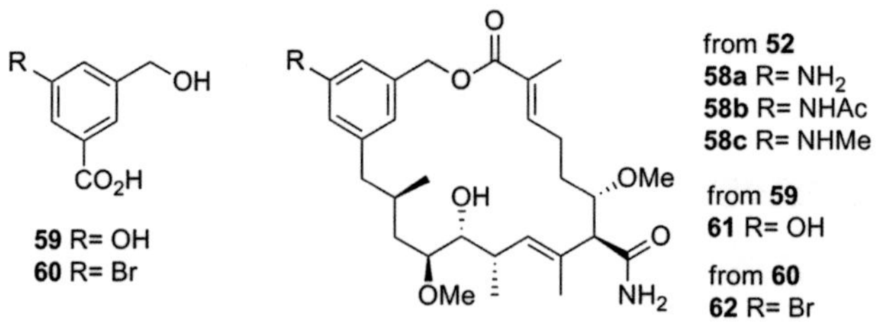

Figure 6. Formation of 20-membered macrolactones after feeding benzyl alcohols 52, 59 and 60 to a AHBA blocked mutant of *S. hygroscopicus*.

CONCLUSIONS

The new term synthetic biology covers an emerging research field that combines the investigative nature of biology with the constructive nature of engineering. Efforts in synthetic biology have largely focused on the creation and perfection of genetic devices and small modules that are constructed from these devices. Artemisinic acid, that was presented as a show case for the preparation of small molecules, is an illustrative example, where the latest approaches and genetic techniques of metabolic engineering will lead to in the near future (Scheme 1).

It is also an illustrative example that the combination of metabolic engineering with chemical synthesis holds much larger future prospects, because in fact artemisinin is the active metabolite. Furthermore, chemical synthesis is able to broaden the chemical space of metabolites thus creating larger structural diversity. The present account covers only one aspect of this emerging field in that only the chemical utility of mutant strains that are specifically blocked in one location of the biosynthetic gene cluster is discussed. Many more combinations of chemical synthesis with metabolically engineered "biofactories" can be envisaged, particularly as synthetic biology currently sets out to create more flexible arrangements and combinations of biosynthetic elements. Ideally, this development will create custom made or "programmable" metabolic pathways in an effort to reach similar flexibility as is typical for chemical synthesis but with larger preciseness. Eventually, the preparation of a target molecule will rely on a tool box composed of biological and chemical synthetic methods that will be combined in a highly flexible manner for a given synthetic problem.

Synthesis quo vadis? A CHEMICAL Synthetic Biology is definitely one answer.

ACKNOWLEDGEMENTS

Contributions from Hannover to this field of research were only possible by a group of excellent co-workers specifically by M. Brünjes, G. Dräger, S. Eichner, T. Frenzel, K. Harmrolfs, T. Knobloch, S. Mayer, A. Meyer, F. Taft and B. Thomaszewski and excellence in the NMR department. Biological testing was performed by F. Sasse (Helmholtz center of infectious disease (HZI), Braunschweig, Germany). We are particularly grateful to H. G. Floss (University of Washington, Seattle, USA) for initial and continuous generous support. Finally, we thank the Fonds der Chemischen Industrie and the Deutsche Forschungsgemeinschaft (grant Ki 397/7 – 1 and Ki 397/13 – 1) for financial contributions.

REFERENCES

[1]   Purnick, P.E.M., Weiss, R. (2009) *Nature Rev. Mol. Cell Biol.* **10**:410 – 422. doi: 10.1038/nrm2698.

[2]   Benner, S.A., Sismour, A.M. (2005) *Nature Rev. Gen.* **6**:533 – 543. doi: 10.1038/nrg1637.

[3]   Klayman, D.L. (1985) *Science* **228**:1049 – 1055. doi: 10.1126/science.3887571.

[4]   Sachs, J., Malaney, P. (2002) *Nature* **415**:680 – 685 doi: 10.1038/415680a.

[5]   Ro, D.-K., Paradise, E.M., Oullet, M., Fisher, K.J., Newman, K.L., Ndungu, J.M., Ho, K.A., Eachus, R.A., Ham, T.S., Kirby, J., Chang, M.C.Y., Withers, S.T., Shiba, Y., Sarpong, R., Kaesling, J.D. (2006) *Nature* **440**:940 – 943. doi: 10.1038/nature04640.

[6]   Lévesque, F., Seeberger, P.H. (2012) *Angew. Chem. Int. Ed.* 51:1706 – 1709. doi: 10.1002/anie.201107446.

[7]   Gupta, P.B., Onder, T.T., Jiang, G., Tao, K., Kuperwasser, C., Weinberg, R.A., Lander, E.S. (2009) *Cell* **138**:645 – 659. doi: 10.1016/j.cell.2009.06.034.

[8]   a) Bao, S., Wu, Q., McLendon, R.E., Hao, Y., Shi, Q., Hjelmeland, A.B., Dewhirst, M.W., Bignerm, D.D., Rich, J.N. (2006) *Nature.* **444**:756 – 760. doi: 10.1038/nature05236;
b) Dean, M., Fojo, T., Bates, S. (2005) *Nat. Rev. Cancer* **5**:275 – 284. doi: 10.1038/nrc1590;
c) Diehn, M., Clarke, M.F. (2006) *J. Natl. Cancer Inst.* **98**:1755 – 1757. doi: 10.1093/jnci/djj505;
d) Diehn, M., Cho, R.W., Lobo, N.A., Kalisky, T., Dorie, M.J., Kulp, A.N., Qian, D., Lam, J.S., Ailles, L.E., Wong, M., Joshua, B., Kaplan, M.J., Wapnir, I., Dirbas, D.M., Somlo, G., Garberoglio, C., Paz, B., Shen, J., Lau, S.K., Quake, S.R., Brown, J.M., Weissman, I.L., Clarke, M.F. (2009) *Nature.* **458**:780 – 783. doi: 10.1038/nature07733;
e) Eyler, C.E., Rich, J.N. (2008) *J. Clin. Oncol.* **26**:2839 – 2845. doi: 10.1200/JCO.2007.15.1829;
f) Li, X., Lewis, M.T., Huang, J., Gutierrez, C., Osborne, C.K., Wu, M.F., Hilsenbeck, S.G., Pavlick, A., Zhang, X., Chamness, G.C., Wong, H., Rosen, J., Chang, J.C. (2008) *J. Natl. Cancer Inst.* **100**:672 – 679. doi: 10.1093/jnci/djn123;

g) Woodward, W.A., Chen, M.S., Behbod, F., Alfaro, M.P., Buchholz, T.A., Rosen, J.M. (2007) *Proc. Natl. Acad. Sci. U.S.A.* **104**:618 – 623. doi: 10.1073/pnas.0606599104.

[9]   Shier, W.T., Rinehart, K.L., Jr., Gottlieb, D. (1969) *Proc. Nat. Acad. Sci. U.S.A.* **63**:198 – 204. doi: 10.1073/pnas.63.1.198.

[10]   Rinehart, K.L., Jr. (1977) *Pure Appl. Chem.* **49**:1361 – 1384. doi: 10.1351/pac197749091361.

[11]   a) Weissman, K.J., Leadlay, P.F. (2005) *Nat. Rev. Microbiol.* **3**:925 – 936. doi: 10.1038/nrmicro1287;
b) Walsh, C.T. (2002) *ChemBioChem* **3**:124 – 134. doi: 10.1002/1439-7633(20020301)3:2/3<124::AID-CBIC124>3.0.CO;2-J;
c) Menzella, H.G., Reeves, C.D. (2007) *Curr. Op. Microbiol.* **10**:238 – 245. doi: 10.1016/j.mib.2007.05.005.

[12]   a) Weist, S., Süssmuth, R.D. (2005) *Appl. Microbiol. Biotechnol.* **68**:141 – 150. doi: 10.1007/s00253-005-1891-8;
b) Kirschning, A., Taft, F., Knobloch, T. (2007) *Org. Biomol. Chem.* **5**:3245 – 3259. doi: 10.1039/b709549j;
c) Wu, M.C., Law, B., Wilkinson, B., Micklefield, J. (2012) *Curr. Opin. Biotechnol.* **23**:931 – 940. doi: 10.1016/j.copbio.2012.03.008.

[13]   Hahn, F., Kirschning, A. (2012) *Angew. Chem. Int. Ed.* **51**:4012 – 4022. doi: 10.1002/anie.201207329.

[14]   a) Cassady, J.M., Chan, K.K., Floss, H.G., Leistner, E. (2004) *Chem. Pharm. Bull.* **52**:1 – 26. doi: 10.1248/cpb.52.1;
b) Kirschning, A, Harmrolfs, K. and Knobloch, T. (2008) *C. R. Chim.* **11**:1523 – 1543. doi: 10.1016/j.crci.2008.02.006;
c) Floss, H.G., Yu; T.-W. and Arakawa, K. (2011) *J. Antibiot.* **64**:35 – 44. doi: 10.1038/ja.2010.139.

[15]   a) Erickson, H.K., Widdison, W.C., Mayo, M.F., Whiteman, K., Audette, C., Wilhelm, S.D. and Singh, R. (2010) *Bioconjugate Chem.* **21**:84 – 92. doi: 10.1021/bc900315y,;
b) Senter, P.D. (2009) *Curr. Opin. Chem. Biol.* **13**:235 – 244. doi: 10.1016/j.cbpa.2009.03.023.

[16]   Eichner, S., Hermane, J., Knobloch, T., Kirschning, A. (2012) Drug Discovery from Natural Products: Mutant Manufactures, RSC Drug Discovery Series No. 25, Cambridge, pp 58 – 78.

[17]   a) Workman, P. (2003) *Curr. Cancer Drug Targets* **3**:297 – 300.
       doi: 10.2174/1568009033481868.;
       b) Neckers, L. and Neckers, K. (2005) *Expert Opin. Emerg. Drugs* **10**:137 – 149.
       doi: 10.1517/14728214.10.1.137;
       c) Whitesell, L. and Lindquist, S.L. (2005) *Nat. Rev. Cancer* **5**:761 – 772.
       doi: 10.1038/nrc1716;
       d) Prodromou, C., Roe, S.M., O'Brien, R., Ladbury, J.E., Piper, P.W. and Pearl, L.H.
       (1997) *Cell* **90**:65 – 75.
       doi: 10.1016/S0092-8674(00)80314-1.

[18]   Biamonte, M.A., van de Water, R., Arndt, J.W., Scannevin, R.H., Perret, D., Lee, W.-
       C. (2010) *J. Med. Chem.* **53**:3 – 17.
       doi: 10.1021/jm9004708.

[19]   Janin, Y.L. (2005) *J. Med. Chem.* **48**:7503 – 7512.
       doi: 10.1021/jm050759r.

[20]   Kim, W., Lee, J.S., Lee, D., Cai, X.F., Shin, J.C, Lee, K., Lee, C.-H., Ryu, S., Paik,
       S.-G., Lee, J.J. and Hong, Y.-S. (2007) *ChemBioChem* **8**:1491 – 1494.
       doi: 10.1002/cbic.200700196.

[21]   PCT Int. Appl. 2008, p 53, Pub. No.: US 2008/0188450 A1.

[22]   Menzella, H.G., Tran, T.-T., Carney, J.R., Lau-Wee, J., Galazzo, J., Reeves, C.D.,
       Carreras, C., Mukadam, S., Eng, S., Zhong, Z., Timmermans, P.B.M.W.M., Murli, S.
       and Ashley, G.W. (2009) *J. Med. Chem.* **52**:1518 – 1521.
       doi: 10.1021/jm900012a.

[23]   Eichner, S., Floss, H.G., Sasse, F., Kirschning, A. (2009) *ChemBioChem* **10**:1801 –
       1805.
       doi: 10.1002/cbic.200900246.

[24]   Eichner, S., Eichner, T., Floss, H.G., Fohrer, J., Hofer, E., Sasse, F., Zeilinger, C.,
       Kirschning, A. (2012) *J. Am. Chem. Soc.* **134**:1673 – 1679.
       doi: 10.1021/ja2087147.

[25]   Taft, F., Brünjes, M., Floss, H.G., Czempinski, N., Grond, S., Sasse, F., Kirschning,
       A. (2008) *ChemBioChem* **9**:1057 – 1060.
       doi: 10.1002/cbic.200700742.

[26]   Kubota, T., Brünjes, M., Frenzel, T., Xu, J., Kirschning, A., Floss, H.G. (2006)
       ChemBioChem **7**:1221 – 1225.
       doi: 10.1002/cbic.200500506.

[27]   Eichner, S., Knobloch, T., Floss,H.G., Fohrer, J., Harmrolfs, K., Hermane, J., Schulz, A., Sasse, F., Spiteller, P., Taft, F., Kirschning, A. (2012) *Angew. Chem.* **124**:776 – 781; *Angew. Chem. Int. Ed.* **51**:752 – 757.
         doi: 10.1002/anie.201106249.

[28]   Taft, F., Brünjes, M., Knobloch, T., Floss, H.G., Kirschning, A. (2009) *J. Am. Chem. Soc.* **131**:3812 – 3813.
         doi: 10.1021/ja8088923.

[29]   Meyer, A., Brünjes, M., Taft, F., Frenzel, T., Sasse, F., Kirschning, A. (2007) *Org. Lett.* **9**:1489 – 1492.
         doi: 10.1021/ol0702270.

[30]   Spiteller, P., Bai, L., Shang, G., Carroll, B.J., Yu, T.-W., Floss, H.G. (2003) *J. Am. Chem. Soc.* **125**:14236 – 14237.
         doi: 10.1021/ja038166y.

[31]   Issell, B.F., Crooke, S.T. (1978) *Cancer Treat. Rev.* **5**:199 – 207.
         doi: 10.1016/S0305-7372(78)80014-0.

[32]   a) Wu, A.M., Senter, P.D. (2005) *Nature Biotech.* **23**:1137 – 1146.
         doi: 10.1038/nbt1141;
         b) Chari, R.V. (2008) *Acc. Chem. Res.* **41**:98 – 107. doi: 10.1021/ar700108g;
         c) Alley, S.C., Okeley, N.M., Senter, P.D. (2010) *Curr. Opin. Chem. Biol.* **14**:529 – 537. doi: 10.1016/j.cbpa.2010.06.170.

[33]   Low, P.S., Henne, W.A., Doorneweerd, D.D. (2008) *Acc. Chem. Res.* **41**:120 – 129.
         doi: 10.1021/ar7000815.

[34]   a) Arap, W., Pasqualini, R., Ruoslahti, E. (1998) *Science* **279**:377 – 380.
         doi: 10.1126/science.279.5349.377.;
         b) Janssen, M.L., Oyen, W.J., Dijkgraaf, I., Massuger, L.F., Frielink, C., Edwards, D.S., Rajopadhye, M., Boonstra, H., Corstens, F.H., Boerman, O.C. (2002) *Cancer Res.* **62**:6146 – 6151;
         c) Haubner, R., Wester, H.J., Burkhart, F., Senekowitsch-Schmidtke, R., Weber, W., Goodman, S.L., Kessler, H., Schwaiger, M. (2001) *Bioconjug. Chem.* **12**:84 – 91.
         doi: 10.1021/bc000071n.

[35]   a) Ladino, C.A., Chari, R.V.J., Bourret, L.A., Kedersha, N.L., Goldmacher, V.S. (1997) *Int. J. Cancer* **73**:859 – 864.
         doi: 10.1002/(SICI)1097-0215(19971210)73:6<859::AID-IJC16>3.0.CO;2-#;
         b) Niculescu-Duvaz, I. (2010) *Curr Opin Mol Ther.* **12**:350 – 360;
         c) Junutula, J.R., Flagella, K.M., Graham, R.A., Parsons, K.L., Ha, E., Raab, H., Bhakta, S., Nguyen, T., Dugger, D.L., Li, G., Mai, E., Lewis Phillips, G.D., Hiraragi,

H., Fuji, R.N., Tibbitts, J., Vandlen, R., Spencer, S.D., Scheller, R.H., Polakis, P., Sliwkowski, M.X. (2010) *Clin. Cancer Res.* **16**:4769 – 4778.
doi: 10.1158/1078-0432.CCR-10-0987.

[36]   Reddy, J.A., Westrick, E., Santhapuram, H.K., Howard, S.J., Miller, M.L., Vetzel, M., Vlahov, I., Chari, R.V., Goldmacher, V.S., Leamon, C.P. (2007) *Cancer Res.* **67**:6376 – 82.
doi: 10.1158/0008-5472.CAN-06-3894.

[37]   Low, P.S., Antony, A.C. (2004) *Adv. Drug Delivery Rev.* **56**:1055 – 1231.
doi: 10.1016/j.addr.2004.02.003.

[38]   Lee, R.J., Wang, S., Low, P.S. (1996) *Biochim. Biophys. Acta* **1312**:237 – 242.
doi: 10.1016/0167-4889(96)00041-9.

[39]   Fu, G.C. (2008) *Acc. Chem. Res.* **41**:1555 – 1564.
doi: 10.1021/ar800148f.

[40]   Further *in vivo* methylation or acylation of the thiol group cannot be excluded.

NANOPARTICLE SUPERLATTICE ENGINEERING WITH DNA

ROBERT J. MACFARLANE* AND CHAD A. MIRKIN

Department of Chemistry, Northwestern University, 2145 Sheridan Rd.,
Evanston, IL 60208, U.S.A.

E-MAIL: *rmacfarl@u.northwestern.edu

Received: 30th January 2013 / Published: 13th December 2013

A major challenge in materials synthesis is developing methodologies to synthesize materials by design, the concept that one can know what building blocks are necessary to create a material *a priori* to synthesis. Typically, the building blocks used to synthesize these materials are atoms or molecules, and the identity of the structure being assembled is a function of which atoms are used and how those atoms are bonded to one another. However, developing materials by design using atoms as building blocks is a significant challenge, as the complex factors that dictate how atoms interact with one another makes predicting what structure will be created from a given set of building blocks a difficult task. Moreover, programmability of these interactions is impossible due to the fact that certain factors (such as electronegativity or atomic number) are immutable for atoms. Therefore, once a given set of atoms is chosen, the resulting structures that can be created are inherently linked to the set of building blocks being used. Linus Pauling famously developed a set of rules that explain (in the context of ionic solids) why certain lattices are preferred over others [1]. However, these rules are really a look backwards, as they lack true predictive power and do not always apply to all ionic solids. Given the challenges associated with using atomic or molecular species as building blocks to create materials by design, a more amenable strategy would be to choose building blocks that are more controllable.

In principle, nanoparticles should offer a much better alternative for synthesizing designer materials, due to the large number of factors that can be manipulated in each building block. For example, one can synthesize different sets of nanoparticles that are identical in terms of their elemental composition, but vary in size or shape [2 – 8]; alternatively, one can synthesize two nanoparticles that are structurally identical at the nanometer scale, but are composed of different atomic constituents [4, 9 – 16]. This versatility provides significantly greater programmability in building block design, and, subsequently, in materials synthesis. However, the challenge associated with using nanoparticles as building blocks is that nanoparticles do not inherently have components that allow them to bond to one another in the manner that electrons allow for atomic bonds to be created. Therefore, to assemble nanoparticles into hierarchical structures, ligands must be attached to the surface of the nanoparticles to dictate how nanoparticles interact with each other [17–22]. Many different ligands have been utilized to achieve this, taking control of ionic [20], van der Waals [21–22], or biological recognition interactions [17, 18] to dictate nanoparticle bonding patterns. An ideal ligand choice to achieve these types of interactions is DNA. DNA is a linear polymer consisting of a series of nucleobases, where the length of the DNA strand is dictated by the number of nucleobases, and interactions between different DNA strands are dictated by the sequence of those bases – two DNA strands will only hybridize to one another if they contain complementary base sequences. DNA is readily synthesized via automated procedures, and can be readily modified with many different functional groups, enabling its attachment to a wide variety of nanoparticle compositions [23–27].

In 1996, the Mirkin group developed the concept of the polyvalent-DNA nanoparticle conjugate, now referred to as the spherical nucleic acid nanoparticle conjugate (SNA-NP). This structure consists of a nanoparticle core, functionalized with a dense monolayer of synthetic oligonucleotides. SNA-NPs have been synthesized with a range of different nanoparticle core compositions, including noble metals (Au [18], Ag [25]), semiconductors (CdSe [28]), oxides (SiO_2 [27], Fe_3O_4 [23]), and polymeric materials [29]. They possess both properties associated with their inorganic core, such as plasmon resonances in the case of noble metals, and the programmable recognition capabilities of DNA. Importantly, they also possess emergent properties that are a direct result of attaching a dense layer of DNA strands to the surface of a nano-object, including enhanced binding strengths [30], innate cellular uptake [31], and greater discrimination against nucleobase sequence mismatch [32]. These enhanced properties have enabled their use in multiple applications, including low limit of detection diagnostics [33] and gene regulation [34].

Since the development of the SNA-NP conjugate in 1996, the Mirkin group, along with others, have taken major strides in being able to utilize this structure as a building block in the synthesis of ordered arrays of nanoparticles. Although early attempts yielded materials that exhibited no long range ordering, short-range ordering and interparticle distance control was demonstrated in subsequent years [35]. Long-range ordering and crystalline materials were achieved for the first time in 2008 by altering the SNA-NP design to utilize DNA linker strands to connect the nanoparticles [36]. These linkers consisted of three basic regions: a recognition sequence complementary to the DNA sequence physically attached to the nanoparticle, a spacer sequence used to control the length of the DNA linker, and a short sticky end presented at the hydrodynamic radius of the SNA-NP. Hybridization interactions between sticky ends on adjacent SNA-NPs allow the nanoparticle to link to one another. Because each SNA-NP can be attached to tens to hundreds of DNA linker strands (depending on the nanoparticle core size) [24], each nanoparticle can form tens to hundreds of DNA connections to adjacent particles.

It is therefore the key hypothesis of our efforts in SNA-NP assembly that, because the driving force for crystallization is the formation of duplex DNA linkages:

The most stable structure will maximize the number of duplex DNA connections between particles.

The simplicity of this hypothesis allows us to understand the behavior of the SNA-NPs in the context of assembly so well that we have been able to develop a set of design rules for the process of crystal formation [17]. Like Pauling's rules for atoms, these rules explain the relative stability of different lattices as a function of their DNA and nanoparticle building blocks. However, unlike Pauling's rules, this rule set allows one to predict the stability of crystal structures before they are ever synthesized. Therefore, one can direct the formation of a desired crystal structure by controlling different aspects of the building blocks, such as nanoparticle size, nanoparticle shape, DNA length, or DNA sequence [17].

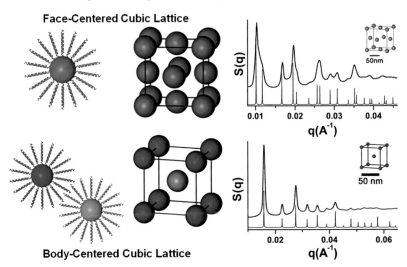

Figure 1. Face-centered cubic lattices are obtained from DNA with self-complementary sticky end sequences, and body-centered cubic lattices are obtained in a binary system where two NPs are functionalized with DNA that contain complementary sticky ends; all lattices are confirmed via small angle X-ray scattering. Figure adapted from ref. 17 (reprinted with permission from publisher).

The first rule in DNA-mediated assembly is: SNA-NPs of equal hydrodynamic radii will form an fcc lattice when using self-complementary DNA sequences, and a bcc lattice when using two SNA-NPs with complementary DNA sequences (Fig. 1). When assembling particles with a self-complementary sticky end, all particles are able to form DNA linkages with all other particles in solution. Therefore, the maximum number of DNA duplexes are formed when each particle is surrounded by the largest number of nearest neighbors, which is achieved when the particles are arranged in a face-centered cubic lattice. This is because fcc lattices represent the densest packing of spheres of a single size. However, when two different sets of particles are combined with sticky ends that are complementary to each other, the maximum number of DNA duplexes is achieved with a body centered cubic arrangement. This is due to the fact that, while a bcc arrangement does not maximize the total number of nearest neighbors for each particle, it does maximize the number of complementary nearest neighbors – the number of nearest neighbors to which a particle can actually form a connection.

To achieve these lattices, particles are combined with the appropriate DNA linkers, then allowed to aggregate. The aggregate (which is initially a disordered, kinetic structure) is annealed, and the thermal energy allows the particles to reposition themselves within the lattice to form an ordered structure. The identity of the lattice is confirmed using small angle X-ray scattering (SAXS), as well as transmission electron microscopy (TEM) (Fig. 2).

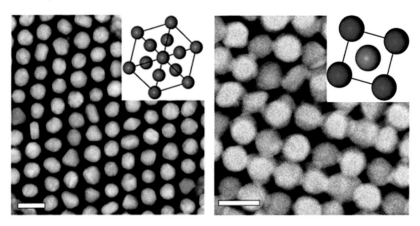

Figure 2. TEM images of fcc (left) and bcc (right) lattices, along with models showing the corresponding lattice orientation (insets). Scale bars are 50 nm. Figure adapted from ref. 17 (reprinted with permission from publisher).

An important aspect of this rule is that it applies to many different particle sizes and DNA lengths. Nanoparticles between 5 and 80 nm have been crystallized into fcc and bcc lattices, where the lattice parameters were controlled to be between 25 and 225 nm [17, 37 – 38]. Importantly, because the nanoparticle size and DNA length are each independent variables, either one (or both) can be used to control the lattice parameters and interparticle distances within a lattice. Thus, nanoparticles can be assembled in a lattice where the unit cell edge length is not dictated by the size of the particles being assembled.

In addition to the large amount of programmability afforded in dictating lattice parameters, this DNA-mediated assembly process also enables a high degree of precision in particle placement. By examining the interparticle distance for each of the fcc and bcc lattices synthesized, the DNA linker rise per base pair (bp) value can be calculated. This value is the increase in interparticle distance afforded by making a DNA strand one basepair (bp) longer, and is calculated to be ~0.255 nm/bp for all fcc and bcc lattices synthesized. This indicates that the DNA-mediated assembly process affords nanometer-scale precision in dictating the lattice parameters of a crystal, as each additional base in a DNA linker adds less than one nanometer to the interparticle distance in the crystal.

In addition to controlling the length of the DNA linkages between particles, it is also possible to control the strength of an individual DNA sticky end duplex, which allows for kinetic products to be accessed. A corollary to the first design rule is therefore that, for two lattices of similar stability, kinetic products can be produced by slowing the rate at which individual DNA linkers de- and subsequently re-hybridize. In order for a lattice to transition from an initial kinetic product to the thermodynamically most favorable state, DNA linkages that tether a nanoparticle in place must be broken, thereby allowing the nanoparticle to reposition itself within an aggregate, where new DNA connections can be formed [39]. Each DNA de- and subsequent re-hybridization event occurs on the timescale of microseconds; this is due to the weak nature of individual sticky end duplexes (which allows DNA duplexes to be rapidly broken) and the high local concentration of DNA (which allows the DNA strands, once separated, to find and hybridize to another adjacent DNA sticky end). By using temperature to control the rate at which these hybridization and dehybridization events occur, one can lock in kinetic products that exist as metastable states. For example, a hexagonal close-packed (hcp) arrangement positions individual nanoparticles to have the same number of nearest neighbors as a particle in an fcc lattice (Fig. 3).

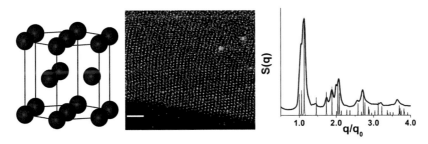

Figure 3. HCP lattices can be obtained as kinetic products by controlling the reorganization rate of particles within a lattice. Scale bar is 100 nm. Figure adapted from ref. 17 (reprinted with permission from publisher).

HCP lattices are only observed as kinetic products in these systems, due to a slight favorability in the energetics of fcc lattices [36, 40]. However, hcp lattices can be observed as kinetic products at early time points in the formation of fcc lattices, as their simpler packing structure enables them to be more readily observed in small clusters of particles. These hcp lattices can be stabilized by slowing the rate of reorganization (i. e., slowing the rate at which DNA bonds are formed and/or broken during the crystallization process). This promotes the growth of hcp clusters over their reorganization to the more favored fcc lattice. This highlights the high degree of control afforded by using DNA strands to assemble nanoparticles.

The distance between two nanoparticles in lattices that are linked with DNA is determined as the sum of the inorganic nanoparticle radii and the length of the DNA linker strands. However, two SNA-NPs that are bound together only interact at their hydrodynamic radius, meaning that there is no direct interaction between the inorganic cores. This allows one to separately control the radius of an inorganic nanoparticle and the overall hydrodynamic radius of an SNA-NP.

Therefore, the second rule of DNA-programmed nanoparticle assembly is:

The overall hydrodynamic radius of a SNA-NP, rather than the sizes of its individual NP or oligonucleotide components, dictates its assembly and packing behavior.

In other words, two nanoparticles can be made to behave as if they were exactly the same size, even if they possess inorganic nanoparticle cores of widely different diameters; this is achieved by altering the DNA linker lengths on the two particles so that the sums of the DNA length and nanoparticle core are equivalent (Fig. 4). This enables the synthesis of CsCl-type lattices, which exhibit the same connectivity as bcc lattices, but utilize particles of with different inorganic cores. Interestingly, appropriate choice of DNA linker lengths allows for two particles to have the same hydrodynamic radius (and therefore same assembly behavior) even if one particle has an inorganic core that is three times the size of the other.

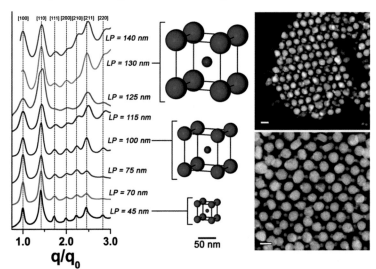

Figure 4. The hydrodynamic radius of a particle dictates its assembly parameters; this can be used to generate CsCl-type lattices. Scale bars are 100 nm. Figure adapted from ref. 17 (reprinted with permission from publisher).

The third rule for DNA-programmed assembly is:

In a binary system based upon complementary SNA-NPs, favored products will tend to have equivalent numbers of each complementary DNA sequence, evenly spaced throughout a unit cell.

In order to achieve more complex crystallographic symmetries, DNA and nanoparticle combinations must be used where, unlike the fcc, bcc, and CsCl lattices mentioned previously, nanoparticle assembly behavior is not equivalent for the two particles in a binary system. This is most readily achieved by changing the ratio of the hydrodynamic sizes of the particles, or by changing the relative number of DNA strands on each particle. The size ratio affects the packing of the SNA-NPs by controlling where particles are positioned relative to one another. For example, a particle with a large hydrodynamic radius can be surrounded by a large number of particles with a relatively small hydrodynamic radius, but each small particle can only be surrounding by a few larger particles; one would therefore predict that the most favored unit cell for this pairing would have a particle stoichiometry that favors the presence of a larger number of the smaller particles. The ratio of the number of DNA strands on each particle type has a different, but equally important effect. Because the driving force for crystallization is the maximization of the number of DNA linkages formed, it is reasonable to assume that a favored unit cell will have roughly equivalent numbers of DNA linkers of complementary types. In other words, if, in a binary system, one particle has twice the number of DNA linkers attached to it, the most favored unit cell will most likely have half as many of that type of particle, so that there are equal numbers of DNA linkers of opposite types. By manipulating these two variables (hydrodynamic size ratio and linker number ratio), three additional crystal structures can be synthesized, isostructural with AlB_2, Cr_3Si, and the alkali-fullerene complex Cs_6C_{60} (Fig. 5). Each of these lattice can be synthesized with varied lattice parameters and particle sizes, in the same manner as the fcc, bcc, and CsCl structures.

Interestingly, when comparing the binary lattices synthesized with this method, a trend can be observed that given crystal symmetries tend to favor specific values of hydrodynamic size ratio and/or DNA linker number ratio.

Thus, rule four is:

Two systems with the same size ratio and DNA linker ratio exhibit the same thermodynamic product.

The importance of this rule is that, once a crystal structure has been synthesized, it allows one to "dial in" a crystal structure using different particle sizes or DNA lengths, simply by tailoring the size and linker ratios to match those of the initial crystal structure. Perhaps more important is the implication that, if these two variables can be used to determine which

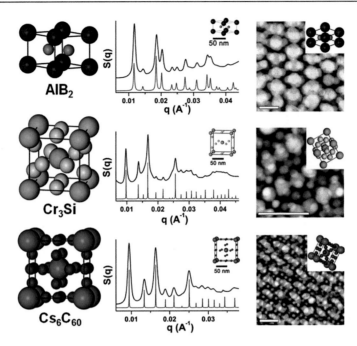

Figure 5. Controlling particle size ratio and DNA linker ratio in a binary system enable the synthesis of lattices with different particle stoichiometries. Scale bars are 50 nm. Figure adapted from ref. 17 (reprinted with permission from publisher).

crystal structure is most stable, one should be able to develop a phase diagram with these two variables as axes. Although it is not possible to directly measure the number of DNA duplexes formed in a given structure (nor to measure the number of DNA duplexes in an unstable structure that does not form), it is possible to develop computational methods that predict these values as a function of nanoparticle size, DNA length, and crystal symmetry. To achieve this, a mean-field model was developed based on established properties of both DNA (such as rise per base pair and persistence length [38]) and SNA-NPs (such as the number of DNA strands per particle [24]). This model was then used to calculate the number of DNA duplexes formed as a function of size ratio, linker ratio, and crystal symmetry, and to construct a phase diagram that predicted which crystal symmetry maximized DNA hybridization interactions for a given size and linker ratio combination (Fig. 6).

It is important to note that many different crystal symmetries were examined, and, for each region of the phase diagram, the most stable symmetry was always found to be one of the four crystal structures obtained experimentally. Moreover, a comparison of the modeled phase diagram and the experimental data demonstrated greater than 90% agreement between the two; i.e. over 90% of the crystal structures obtained experimentally were correctly predicted by the phase diagram as being the most stable arrangement of particles.

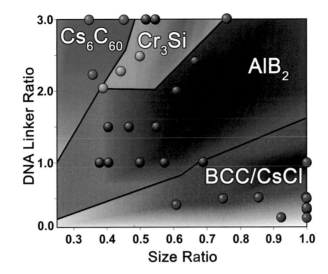

Figure 6. Phase diagram relating experimentally controllable parameters to predicted crystal structure. Experimental data points are shown to correlate with theoretical phase diagram and are color coded to the experimentally observed structure. Figure adapted from ref. 17 (reprinted with permission from publisher).

Because each of the SNA-NPs used to assemble these superlattices is functionalized with a large number of DNA strands, it is possible to put multiple DNA sequences on a single nanoparticle.

Therefore, the fifth rule of DNA-programmed assembly is:

SNA-NPs can be functionalized with more than one oligonucleotide bonding element, providing access to crystal structures not possible with single element SNA-NPs.

For example, a SNA-NP that is functionalized with linkers containing self-complementary sticky ends would be predicted to form an fcc lattice as its thermodynamic product, as per rule 1. However, if this SNA-NP were also functionalized with a second type of DNA linker that contained a sticky end complementary to the sticky ends of a different SNA-NP, additional DNA linkages could be created. This enables the synthesis of NaCl-type lattices, where the self-complementary sticky ends dictate the formation of an fcc lattice, but the additional sticky ends allow a second nanoparticle to fill into the octahedral holes in the fcc lattice while it forms (Fig. 7).

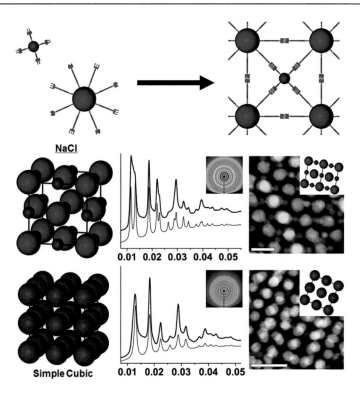

Figure 7. By hybridizing multiple types of linkers to a single SNA-NP, multiple types of DNA bonds can be generated in a single lattice. Scale bars are 50nm. Figure adapted from ref. 17 (reprinted with permission from publisher).

It should be noted, however, that the hydrodynamic size ratio of the two particles must be ~0.40 in order to form an NaCl lattice. If one of the particles has a hydrodynamic radius that is too large or too small, the sticky ends will not be positioned at the correct location to enable both the self- and non-self-complementary linkages to be formed at the same time. However, as mentioned in the second rule, the inorganic nanoparticle core size can be varied separately from the hydrodynamic radius. When assembling a NaCl-type lattice where the hydrodynamic size ratio is still ~0.40, but the inorganic particle core sizes are equal, it is actually a simple cubic lattice that is formed (as defined by the positions of the inorganic cores) (Fig. 7).

The crystal symmetry for each of these lattices is defined by the positions and identities of the inorganic cores. This enables a strategy to increase the complexity of lattices that can be synthesized by using "hollow" (or coreless) particles and therefore the sixth rule is:

The crystal symmetry of a lattice is dictated by the position of the inorganic cores; an SNA with no inorganic core can be used to "delete" a particle at a specified position within a unit cell.

These particles without inorganic cores are achieved by first functionalizing a gold particle with DNA, then cross-linking the DNA strands at the particle surface [29]. The gold particle can then be dissolved and, because the DNA strands are covalently attached to one another, the SNA retains its overall structure and binding properties [41]. When these SNAs with no inorganic core are used to assemble superlattices, they act as placeholders in the assembly process – they occupy a specific site within a unit cell and stabilize the lattice, but do not actually contribute to the arrangement of particles that defined the crystal symmetry. As an example, if one uses a hollow particle in place of one of the SNA-NPs in a bcc lattice, the bcc lattice becomes a simple cubic arrangement (Fig. 8). When using an AlB₂-type structure as a template, this strategy enables the synthesis of either simple hexagonal or graphite-like lattices when the hollow particle has the smaller or larger hydrodynamic radius, respectively (Fig. 8).

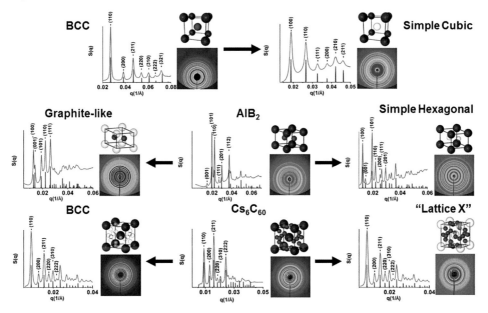

Figure 8. Utilizing a SNA with no inorganic core enables one to synthesize lattices where a particular nanoparticle type has been "deleted" from the unit cell. Note that the coreless SNAs are synthesized prior to superlattice formation, as opposed to being removed post-superlattice assembly. Figure adapted from ref. 41 (reprinted with permission from publisher).

If one assembles a Cs_6C_{60}-type lattice with hollow particles, it transforms to either a bcc lattice (when the smaller particle has a hollow core), or a lattice with no known mineral equivalent (when the larger particle has a hollow core). Because this lattice has not been constructed in any other method to date, it has no name and is therefore referred to as "lattice X". (Fig. 8) The construction of this symmetry highlights the complexity with which one can program crystallographic symmetry using DNA and nanoparticles as building blocks.

Each of the superlattices discussed so far have utilized spherical particles as building blocks, which limits them to inherently isotropic interactions with adjacent particles. However, if one utilizes particles of different shapes, these different geometries can be used to direct DNA interactions between particles along specific vectors [42]. This is because DNA hybridization will be maximized with particles align themselves to have their largest facets facing one another.

The seventh rule in DNA-directed particle assembly is thus:

Nanoparticle superlattices based upon anisotropic particles with flat faces will assemble into a lattice that maximizes the amount of parallel, face-to-face interactions between particles.

When DNA strands are attached to a planar surface, they all tend to point in the same direction in order to minimize repulsive interactions between adjacent strands, and thus the sticky ends are all positioned at a specific distance away from the particle surface. This means that, when two particles with flat faces approach one another, a significantly larger number of DNA connections will be made when those two particles are aligned with their flat faces parallel to one another, as this will position all of the DNA strands on those faces such that they can engage in hybridization. There is therefore a huge thermodynamic preferences for aligning anisotropic particles in this manner, and this can be used to dictate crystal symmetry [43]. When utilizing flat triangular prisms (thickness of ~7 nm, triangular edge length variable between 40 and 150 nm), which are essentially planar structures, a one-dimensional lamellar stack of prisms is obtained, where all of the prisms are aligned face to face (Fig. 9). Alternatively, rod-like structures arrange themselves in a hexagonal array, as this maximizes DNA interactions between the long axes of the rods (Fig. 9). Rhombic dodecahedra (a twelve-sided polyhedron that naturally packs into an fcc lattice) forms the predicted fcc structure, but importantly, each particle in the lattice has both positional and rotational ordering. This means that each particle within the lattice is aligned with their flat faces parallel to one another (Fig. 9). Similarly, octahedra (eight-sided polyhedral) adopt a bcc conformation with positional and rotational order, where each particle is surrounded by 8 nearest neighbors (Fig. 9).

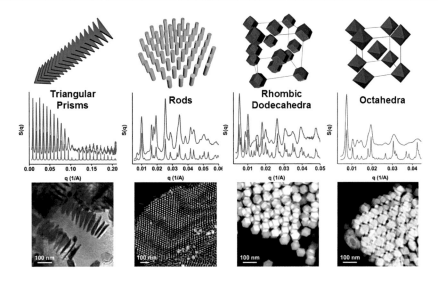

Figure 9. Anisotropic particles form superlattices that maximize the amount of parallel, face-to-face interactions between adjacent particles, as this maximizes the possibility for DNA hybridization events to occur. Figure adapted from ref. 42 (reprinted with permission from publisher).

CONCLUSIONS

In this work, a series of design rules has been presented that enable one to utilize programmable DNA interactions to generate a large number of nanoparticle superlattices. These rules provide access to a readily manipulated design space, where factors such as crystallographic symmetry, lattice parameters, and particle sizes can be independently controlled. This enables the synthesis of a wide variety of crystal structures that cannot be generated through other assembly techniques. These superlattices can be synthesized from particles of over an order of magnitude of different sizes, and the lattice parameters can be tuned over a few hundred nanometers. Importantly, these rules allow one to independently control factors such as interparticle distance, crystal symmetry, and particle identity. Because these factors can be used to influence the emergent properties of nanoparticle superlattices, this methodology should prove useful in generating designer materials with programmable physical properties.

Future directions will include the synthesis of nanoparticles of different material compositions, as well as the experimental investigation of unique plasmonic, photonic, magnetic, or catalytic properties of these unique structures [4, 12, 22, 44 – 46].

REFERENCES

[1] Pauling, L. (1960) The nature of the chemical bond and the structure of molecules and crystals; an introduction to modern structural chemistry, 3rd ed., Cornell University Press, Ithaca, N.Y.

[2] Brus, L.E. (1983) *Journal of Chemical Physics* **79**:5566 – 5571.
doi: 10.1063/1.445676.

[3] Burda, C., Chen, X.B., Narayanan, R., El-Sayed, M.A. (2005) *Chemical Reviews* **105**:1025 – 1102.
doi: 10.1021/cr030063a.

[4] Daniel, M.C., Astruc, D. (2004) *Chemical Reviews* **104**:293 – 346.
doi: 10.1021/cr030698+.

[5] Kelly, K.L., Coronado, E., Zhao, L.L., Schatz, G.C. (2003) *Journal of Physical Chemistry B* **107**:668 – 677.
doi: 10.1021/jp026731y.

[6] Link, S., El-Sayed, M.A. (1999) *Journal of Physical Chemistry B* **103**:8410 – 8426.
doi: 10.1021/jp9917648.

[7] Rossetti, R., Nakahara, S., Brus, L.E. (1983) *Journal of Chemical Physics* **79**:1086 – 1088.
doi: 10.1063/1.445834.

[8] Xia, Y.N., Yang, P.D., Sun, Y.G., Wu, Y.Y., Mayers, B., Gates, B., Yin, Y.D., Kim, F., Yan, Y.Q. (2003) *Advanced Materials* **15**:353 – 389.
doi: 10.1002/adma.200390087.

[9] Ajayan, P.M. (1999) *Chemical Reviews* **99**:1787 – 1799.
doi: 10.1021/cr970102g.

[10] Alivisatos, A.P. (1996) *Science* **271**:933 – 937.
doi: 10.1126/science.271.5251.933.

[11] Fernandez-Garcia, M., Martinez-Arias, A., Hanson, J.C., Rodriguez, J.A. (2004) *Chemical Reviews* **104**:4063 – 4104.
doi: 10.1021/cr030032f.

[12] Law, M., Goldberger, J., Yang, P.D. (2004) *Annual Review of Materials Research* **34**:83 – 122.
doi: 10.1146/annurev.matsci.34.040203.112300.

[13] Oh, M., Mirkin, C.A. (2005) *Nature* **438**:651 – 654.
doi: 10.1038/nature04191.

[14]   Spokoyny, A.M., Kim, D., Sumrein, A., Mirkin, C.A. (2009) *Chemical Society Reviews* **38**:1218 – 1227.
doi: 10.1039/b807085g.

[15]   Trindade, T., O'Brien, P., Pickett, N.L. (2001) *Chemistry of Materials* **13**:3843 – 3858.
doi: 10.1021/cm000843p.

[16]   Xia, Y.N., Xiong, Y.J., Lim, B., Skrabalak, S.E. (2009) *Angewandte Chemie – International Edition* **48**:60 – 103.
doi: 10.1002/anie.200802248.

[17]   Macfarlane, R.J., Lee, B., Jones, M.R., Harris, N., Schatz, G.C., Mirkin, C.A. (2011) *Science* **334**:204 – 208.
doi: 10.1126/science.1210493.

[18]   Mirkin, C.A., Letsinger, R.L., Mucic, R.C., Storhoff, J.J. (1996) *Nature* **382**:607 – 609.
doi: 10.1038/382607a0.

[19]   Mucic, R.C., Storhoff, J.J., Mirkin, C.A., Letsinger, R.L. (1998) *Journal of the American Chemical Society* **120**:12674 – 12675.
doi: 10.1021/ja982721s.

[20]   Kalsin, A.M., Fialkowski, M., Paszewski, M., Smoukov, S.K., Bishop, K.J.M, Grzybowski, B.A. (2006) *Science* **312**:420 – 424.
doi: 10.1126/science.1125124.

[21]   Shevchenko, E.V., Talapin, D.V., Kotov, N.A., O'Brien, S., Murray, C.B. (2006) *Nature* **439**:55 – 59.
doi: 10.1038/nature04414.

[22]   Talapin, D.V., Lee, J.S., Kovalenko, M.V., Shevchenko, E.V. (2010) *Chemical Reviews* **110**:389 – 458.
doi: 10.1021/cr900137k.

[23]   Cutler, J.I., Zheng, D., Xu, X.Y., Giljohann, D.A., Mirkin, C.A. (2010) *Nano Letters* **10**:1477 – 1480.
doi: 10.1021/nl100477m.

[24]   Hurst, S.J., Lytton-Jean, A.K.R., Mirkin, C.A. (2006) *Analytical Chemistry* **78**:8313 – 8318.
doi: 10.1021/ac0613582.

[25]   Lee, J.S., Lytton-Jean, A.K.R., Hurst, S.J., Mirkin, C.A. (2007) *Nano Letters* **7**:2112 – 2115.
doi: 10.1021/nl071108g.

[26] Xue, C., Chen, X., Hurst, S.J., Mirkin, C.A. (2007) *Advanced Materials* **19**:4071 – 4074.
doi: 10.1002/adma.200701506.

[27] Young, K.L., Scott, A.W., Hao, L.L., Mirkin, S.E., Liu, G.L., Mirkin, C.A. (2012) *Nano Letters* **12**:3867 – 3871.
doi: 10.1021/nl3020846.

[28] Mitchell, G.P., Mirkin, C.A., Letsinger, R.L. (1999) *Journal of the American Chemical Society* **121**:8122 – 8123.
doi: 10.1021/ja991662v.

[29] Cutler, J.I., Zhang, K., Zheng, D., Auyeung, E., Prigodich, A.E., Mirkin, C.A. (2011) *Journal of the American Chemical Society* **133**:9254 – 9257.
doi: 10.1021/ja203375n.

[30] Lytton-Jean, A.K.R., Mirkin, C.A. (2005) *Journal of the American Chemical Society* **127**:12754 – 12755.
doi: 10.1021/ja052255o.

[31] Rosi, N.L., Giljohann, D.A., Thaxton, C.S., Lytton-Jean, A.K.R., Han, M.S., Mirkin, C.A. (2006) *Science* **312**:1027 – 1030.
doi: 10.1126/science.1125559.

[32] Elghanian, R., Storhoff, J.J., Mucic, R.C., Letsinger, R.L., Mirkin, C.A. (1997) *Science* **277**:1078 – 1081.
doi: 10.1126/science.277.5329.1078.

[33] Thaxton, C.S., Elghanian, R., Thomas, A.D., Stoeva, S.I., Lee, J.S., Smith, N.D., Schaeffer, A.J., Klocker, H., Horninger, W., Bartsch, G., Mirkin, C.A. (2009) *Proceedings of the National Academy of Sciences of the U.S.A.* **106**:18437 – 18442.
doi: 10.1073/pnas.0904719106.

[34] Giljohann, D.A., Seferos, D.S., Prigodich, A.E., Patel, P.C., Mirkin, C.A. (2009) *Journal of the American Chemical Society* **131**:2072 – 2073.
doi: 10.1021/ja808719p.

[35] Park, S.J., Lazarides, A.A., Storhoff, J.J., Pesce, L., Mirkin, C.A. (2004) *Journal of Physical Chemistry B* **108**:12375 – 12380.
doi: 10.1021/jp040242b.

[36] Park, S.Y., Lytton-Jean, A.K.R., Lee, B., Weigand, S., Schatz, G.C., Mirkin, C.A. (2008) *Nature* **451**:553 – 556.
doi: 10.1038/nature06508.

[37] Hill, H.D., Macfarlane, R.J., Senesi, A.J., Lee, B., Park, S.Y., Mirkin, C.A. (2008) *Nano Letters* **8**:2341 – 2344.
doi: 10.1021/nl8011787.

[38] Macfarlane, R.J., Jones, M.R., Senesi, A.J., Young, K.L., Lee, B., Wu, J.S., Mirkin, C.A. (2010) *Angewandte Chemie – International Edition* **49**:4589 – 4592.
doi: 10.1002/anie.201000633.

[39] Macfarlane, R.J., Lee, B., Hill, H.D., Senesi, A.J., Seifert, S., Mirkin, C.A. (2009) *Proceedings of the National Academy of Sciences of the U.S.A.* **106**:10493 – 10498.
doi: 10.1073/pnas.0900630106.

[40] Woodcock, L.V. (1997) *Nature* **385**:141 – 143.
doi: 10.1038/385141a0.

[41] Auyeung, E., Cutler, J.I., Macfarlane, R.J., Jones, M.R., Wu, J.S., Liu, G., Zhang, K., Osberg, K.D., Mirkin, C.A. (2012) *Nature Nanotechnology* **7**:24 – 28.
doi: 10.1038/nnano.2011.222.

[42] Jones, M.R., Macfarlane, R.J., Lee, B., Zhang, J.A., Young, K.L., Senesi, A.J., Mirkin, C.A. (2010) *Nature Materials* **9**:913 – 917.
doi: 10.1038/nmat2870.

[43] Jones, M.R., Macfarlane, R.J., Prigodich, A.E., Patel, P.C., Mirkin, C.A. (2011) *Journal of the American Chemical Society* **133**:18865 – 18869.
doi: 10.1021/ja206777k.

[44] Bell, A.T. (2003) *Science* **299**:1688 – 1691.
doi: 10.1126/science.1083671.

[45] Halas, N.J., Lal, S., Chang, W.S., Link, S., Nordlander, P. (2011) *Chemical Reviews* **111**:3913 – 3961.
doi: 10.1021/cr200061k.

[46] Jones, M.R., Osberg, K.D., Macfarlane, R.J., Langille, M.R., Mirkin, C.A. (2011) *Chemical Reviews* **111**:3736 – 3827.
doi: 10.1021/cr1004452.

Designing, Measuring, and Controlling Functional Molecules and Precise Assemblies

Yue Bing Zheng and Paul S. Weiss*

California NanoSystems Institute and Departments of Chemistry & Biochemistry
and Materials Science & Engineering,
University of California, Los Angeles,
CA 90095, U.S.A.

E-Mail: *psw@cnsi.ucla.edu

Received: 5th February 2013 / Published: 13th December 2013

Abstract

Bottom-up assembly of functional molecules on solid surfaces provides a promising approach toward ever smaller and more functional devices. Molecule-substrate and intermolecular interactions can be exploited so that when molecules are transferred from solutions to substrates targeted structures can be obtained. Understanding correlations between the interactions and function of molecules is essential to elucidating the rules of working towards the ultimate limits of miniaturization. This paper covers our recent progress towards this goal by measuring single molecules and precise assemblies confined in self-assembled monolayers (SAMs). By isolating molecules in SAMs, we are able to obtain both accurate measurements and precise control of function. The measurements, in combination with theoretical calculations, allow us to apply molecular design to optimize function and to direct the assembly of molecules. We are now applying the assembly strategies that we have developed for flat surfaces to curved and faceted substrates, while developing new tools to measure the environment, interactions, and dynamics of single molecules and precise assemblies.

INTRODUCTION

Self- and directed assembly of functional molecules on surfaces provide promising approaches towards ever smaller devices and systems featuring superior performance, cost-effectiveness, and low energy consumption [1 – 8]. A broad range of functional molecules has been synthesized and studied in solution where molecules are randomly distributed and cannot be addressed individually [9, 10]. However, transferring molecules from solutions to solid surfaces is required for device applications and detailed measurements [6, 7, 11 – 14]. Molecules function differently on surfaces due to molecule-substrate and intermolecular interactions. This difference makes it important to characterize the effects of interactions on structure and function of surface-bound molecules.

By measuring both single molecules and precise assemblies confined in self-assembled monolayers (SAMs), we are able to obtain accurate information on the structures, interactions, and function of molecules simultaneously at these scales [5 – 7, 11, 15, 16]. Importantly, we measure statistically significant data sets while retaining all of the single-molecule/assembly data so as to be able to elucidate the effects of interactions, environments, measurement conditions, etc. [16 – 19]. Here, we present our recent progress in both directed assembly and measurements. We have developed and applied new imaging and spectroscopic tools for functional measurements [5, 15, 20]. We aim to operate the molecules together in precise assemblies, both cooperatively and hierarchically, in analogy to biological muscles [8]. Our initial efforts in this area reveal both interference and cooperativity. The measurements, in combination with theoretical calculations, allow us to apply molecular design to optimize function and to control assembly [5, 21, 22]. Some of the assembly strategies that we have developed for flat surfaces can be applied to curved and faceted substrates, but new tools must be developed to measure the environment, interactions, and dynamics of single molecules and precise assemblies [23, 24].

ASSEMBLY

We exploit defects in SAMs to insert functional molecules and to control the placement and orientation of these molecules [15, 16, 25 – 31]. Alkanethiolate SAMs on Au substrates are the well-characterized systems that have the advantages of high stability, easy fabrication, and precise tenability [30, 32 – 43]. Scanning tunneling microscopy (STM) provides molecularly resolved images of the SAMs on Au surfaces that reveal several types of defects [5]. Some of the common defects include substrate defects (*i.e.*, step edges and vacancy islands) and monolayer defects (*i.e.*, tilt domain boundaries and regions of poorly ordered molecules) [5]. At these defects, the alkyl chains cannot tightly pack into their crystalline tilted configurations that have maximized van der Waals forces [5]. We have addressed these defects by designing and assembling symmetric upright cage molecules to simplify both assembly and defect structures. As noted above, we exploit defects in SAMs to insert single molecules as well as pairs, lines, and clusters of molecules [7, 16, 25, 26].

Controlled defects in SAMs facilitate insertion of functional molecules by limiting substrate access. Defect type and density can be controlled by molecular design and by processing the films [34, 35, 44]. Multiple methods have been developed for insertion, including solution deposition, vapor deposition, and microcontact insertion printing [16, 30, 45, 46]. Alternatively, in order to isolate single molecules within domains of SAMs, we employ co-adsorption of functional molecules and matrix molecules [27 – 29, 45, 47].

New types of molecules with greater symmetry such as adamantanethiols, carboranethiols, and cubanethiols are being designed and assembled on Au and other substrates to gain new insight into self-assembly and to control defects [44, 48]. The limited degrees of freedom of cage molecules have significant advantages in terms of defect types and densities as compared to linear molecules. These defects will also be exploited for insertion and control of functional molecules.

PHOTOISOMERIZATION

One key challenge has been coupling external energy to molecules on surfaces in order to perform useful work at the nanoscale. When adsorbed directly on metal surfaces, photo-excited molecules are quenched by the underlying substrate and steric hindrance from the proximate molecules. By isolating tethered azobenzene molecules within the domains of a tightly packed SAMs where the functional azobenzene moiety protrudes above the SAMs to reduce both coupling with substrates and interaction with surroundings, we demonstrated reversible *trans-cis* isomerization (Fig. 1) [27].

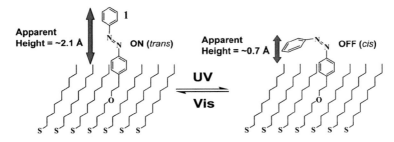

Figure 1. Schematic showing single azobenzene-functionalized molecules (1) isolated in a decanethiolate SAM on Au{111}. Azobenzene undergoes reversible photoisomerization observed as a change in apparent height in STM images. We define trans and cis as ON and OFF states with apparent heights of 2.1 ± 0.3 and 0.7 ± 0.2 Å with respect to the surrounding matrix, respectively. Reproduced with permission from reference 27. Copyright 2008 American Chemical Society.

Scanning tunneling microscopy was employed to image azobenzene molecules and to track their photoisomerization. As shown in Figure 2A, azobenzene molecules were initially in their thermodynamically stable *trans* state and appeared as 2.1 ± 0.3 Å protrusions over the 1-decanethiolate matrix in topographic STM images (also see figure 1). Upon exposure to

UV light (~365 nm; ~12 mW/cm^2), the molecules isomerized to their *cis* state associated with a reduction of the apparent height of ~1.4 Å (figures 2B – 2E). Measurements with STM indicate that with UV illumination for 160 min, over 90 % of the azobenzene molecules isomerized from *trans* to *cis*. Subsequent illumination with visible light (~450 nm; ~6 mW/cm^2) for 30 min switched ~50 % of the molecules back to *trans* conformation (Fig. 2F). The number of molecules switched as a function of UV exposure time is fitted into an exponential curve with a decay constant τ of 54 ± 15 min (Fig. 2G) [27].

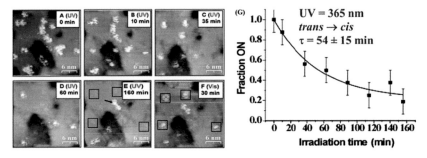

Figure 2. (A-F) Measurements with STM show reversible trans-cis photoisomerization of tethered azobenzene molecules upon irradiation with UV (~365 nm) and visible (~450 nm) light. **(G)** The switching of azobenzene molecules from the more stable trans form as a function of UV illumination time is fitted into exponential function with a decay constant (τ) of 54 ± 15 min. Imaging conditions: $V_{sample} = 1$ V; $I_{tunneling} = 2$ pA. Reproduced with permission from reference 27. Copyright 2008 American Chemical Society.

As an instrument for local surface analysis, STM scans only a small area with few, dilute, single azobenzene molecules isolated in SAM matrices. In order to have statistically significant numbers of measurements of function or dynamics, thousands of STM images must be recorded. This makes it extremely time-consuming to track the time-dependent photophysics or photochemistry that require series of images after every light exposure of the sample. To enable more efficient measurements, we have developed a surface-enhanced Raman spectroscopy (SERS) method to track azobenzene isomerization in the far field [20]. Surface-enhanced Raman spectroscopy, which is based on enhanced interactions of light with molecular vibrations, has played an important role in the study of molecular structure and conformation due to its non-invasiveness, single-molecule sensitivity, and vibrational signatures of chemical identity and bonding changes [41 – 47]. We employed focused ion beam lithography to fabricate cylindrical nanoholes in square arrays on the Au substrates for the enhancement (Fig. 3a). The areas of the substrates surrounding the nanoholes are also accessible to STM and confirm the isolation of single azobenzene molecules in SAMs. Figure 3b shows the Raman spectra of azobenzene recorded at areas with and without nanoholes, respectively. While no obvious Raman signal appears in the spectra recorded from the areas without nanoholes, a strong signal with five peaks was obtained for the azobenzene in the area of nanoholes due to the plasmonic enhancement from the patterned substrate. We assign the five peaks as the P1-P5 modes. Closely coupled theoretical calculations identify the modes and help select which can be used as indicators of switching and which others can used as controls because they are insensitive to isomerization state [20 – 22]. We obtain the switching kinetics (*i.e.*, decay constant τ as defined in figure 2G) of azobenzene by tracking the Raman mode changes as a function of light exposure time. Employment of SERS has accelerated our measurements and thus the optimization of function through molecular design.

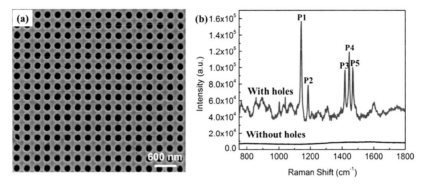

Figure 3. (a) Scanning electron micrograph of a nanohole array on Au{111} on mica fabricated by focused ion beam lithography. **(b)** Raman spectra were recorded from the substrate areas with nanoholes (red) while there were no discernible spectral features from the areas without nanoholes (black). Five vibrational modes are identified as P1-P5. Reproduced with permission from reference 20. Copyright 2011 American Chemical Society.

We have established the role of tether conductivity on the photoisomerization of azobenzene-functionalized molecules on Au surfaces [21]. Molecules were designed so as to tune the conductivity of the tethers that separate the functional azobenzene moiety from the underlying Au substrate (azobenzene with a saturated, non-conductive tether is indicated as **1** in figure 1; azobenzene with conductive tether is indicated as **2** in figure 4).

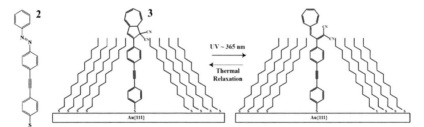

Figure 4. Left: Schematic of azobenzene with a conductive, phenylene ethynylene tether (**2**). Right: Schematic of single dihydroazulene-functionalized molecules (DHA, **3**) in alkanethiol SAMs. The DHA molecules isomerize to vinylheptafulvene (VHF) upon UV irradiation and undergoes back reaction via thermal relaxation. Reproduced with permission from reference 21. Copyright 2012 American Chemical Society. Reproduced with permission from reference 22. Copyright 2013 American Chemical Society.

The decay constants from the SERS analyses reveal that photoisomerization on the Au surface is reduced when the conductivity of the tether is increased (Table 1), consistent with the importance of quenching of the excitation by electronic coupling to the substrate [21].

New molecular switches have been explored to enhance photoswitching efficiency on surfaces. One example is dihydroazulene (DHA, **3** in figure 4). DHA is a photochromic molecule that reversibly switches between two states [22]. We employed SERS to measure the photoreaction kinetics of isolated single dihydroazulene-functionalized molecules in SAMs on Au substrates. The measurements showed that the molecules underwent a ring-opening reaction upon illumination with UV light and switched back to the initial isomer via thermal relaxation. Photokinetic analyses reveal the higher efficiency of the DHA photo-reaction on solid substrates than that of azobenzene isomerization (Table 1) [22].

Table 1. Comparison of SERS-measured decay time constant and photoswitching cross section for isolated single photoswitchable molecules in SAM matrices: azobenzene with non-conductive tether **1**, azobenzene with conductive tether **2**, and dihydroazulene **3**. The smaller time constant and larger cross section indicate higher switching efficiency of DHA as compared to the two tethered azobenzene molecules studied [21, 22].

Photoswitches	τ (min)	σ (cm^2)
Azobenzene 1	38	4.1×10^{-19}
Azobenzene 2	61	2.6×10^{-19}
Dihydroazulene 3	10	1.5×10^{-18}

COORDINATED ACTION

While measuring matrix-isolated single molecules in SAMs provides insight into molecule-substrate interactions and their effects on function, two and more molecules neighboring each other are common in practical applications. To elucidate the interactions and function of the neighboring molecules requires new test beds with precise assemblies of molecules in well-defined environments. While biological systems regularly exploit cooperative action, interference is more commonly observed in synthetic systems studied to date.

By developing new assembly strategies, we have fabricated one-dimensional (1D) chains of tethered azobenzene in SAMs that serve as test beds for measuring and understanding intermolecular interactions and function [49, 50]. Interestingly, photoswitching of the chains exhibits coordinated actions where the molecules within individual chains isomerize coop-eratively (albeit with lower efficiency) compared to the random switching of the isolated single molecules (Fig. 2). The coordinated action of molecules in assemblies is an important step along the way toward linking molecular-scale motion to macroscale function as ex-emplified by biomolecules working hierarchically in nature in order to produce complex functions. For tethered azobenzene chains, we infer that the cooperative action arises partly from electronic coupling along the chain [50]. We found that the thicknesses of the 1D chains play a key role in the efficiency of photoisomerization. Complete reversibility in the isomerization of one-row chains was observed, but the two-row chains had partial switching. Once the molecules were assembled into 2D clusters, the switching efficiency was further reduced. Moreover, the electronic coupling along the chains inspired us to trigger the switching of the chains locally with electrons from the STM tip positioned atop one of the molecules in the chains [49].

BEYOND FLATLAND

Most single-molecule measurements of function have been limited to molecules assembled on atomically flat surfaces that are accessible to surface characterization tools [5]. However, most surfaces in practical applications deviate from ideal atomic flatness [51, 52]. In particular, nanoparticles are emerging as one of the most important functional nanomaterials for such applications as nanoscale catalysts, drug delivery, and theranostics [31]. Nanoparticles have facets, edges, vertices, and curvatures that influence their interactions with molecules and function. Better understanding of the morphological effects of curvature, faceting, and the local chemical environment will enable control of nanoparticle functionalization in ligand exchange in targeted delivery of therapeutics and theranostics and in catalysis [53, 54]. This effort requires development of new assembly strategies and analytical tools that are compatible to the curved surfaces.

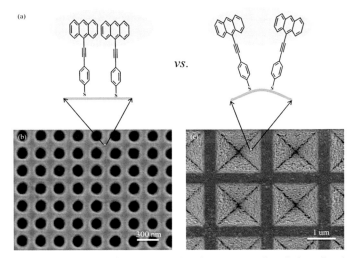

Figure 5. (a) Schematic of UV-exposed 9-(4-mercapto-phenylethynyl)anthracene pairs on atomically flat (left) and curved surfaces (right). Scanning electron micrographs of the two types of surfaces integrated with SERS substrates: **(a)** epitaxial Au on mica patterned with arrays of nanoholes and **(b)** Au nanoparticles on patterned silicon. Reproduced with permission from reference 23. Copyright 2012 American Chemical Society.

In initial efforts along these lines, we assembled pairs of thiolate-linked anthracene phenylethynyl molecules on both flat and curved Au surfaces (Fig. 5). By employing SERS, we identified the UV-induced photoreaction paths of the molecules (Fig. 6) and revealed the effects of nanoscale morphology of substrates on the photoreactions [23]. We found that surface constraint drove the unfavorable [4+4] photoreaction of anthracene rather than [4+2] reaction that is favored in solution where there are no conformational constraints. The SAMs on curved surfaces exhibited dramatically lower regioselective photoreaction kinetics and efficiencies than those on atomically flat surfaces. We infer that the slower reaction kinetics and lower efficiency of the reactions on the curved surfaces arise from the larger intermolecular distances and variable orientations in the SAMs. Moreover, we propose that the surface-dependent properties of such molecules (molecular pairs) can be used to probe the local chemical environment on curved surfaces [31, 55, 56]. New molecular probes with higher sensitivity to curvature and local environment are now being designed and applied.

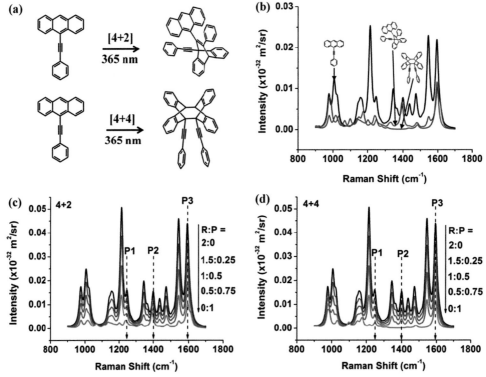

Figure 6. (a) Two photoreaction paths of 9 phenylethynylanthracene in solution: [4+2] and [4+4]. **(b)** Calculated Raman spectra of reactant (R), [4+2] product, and [4+4] product exhibit different vibrational modes. **(c)** A series of simulated Raman spectra with various mole fractions of reactant (R) and [4+2] product (P) (indicated by R:P). **(d)** A series of simulated Raman spectra with various convoluted mole fractions of reactant (R) and [4+4] product (P) (indicated by R:P). Modes P1, P2, and P3 were used in our analyses of the photoreaction paths and kinetics. Reproduced with permission from reference 23. Copyright 2012 American Chemical Society.

Summary and Perspective

In summary, measuring isolated single molecules and precise assemblies confined in well-defined SAMs provides insight into molecule-substrate and intermolecular interactions as well as their effects on function of molecules. These measurements are essential to molecular design in order to optimize interactions and function and to control assembly.

We have achieved reversible photoisomerization of tethered azobenzene as single molecules and 1D chains in SAMs on Au surfaces. Molecules constituting the 1D chains photoisomerized cooperatively but more slowly than did isolated single molecules. Isomerization efficiency is further reduced in 2D assemblies. These first results will guide us in designing new functional molecular assemblies with higher efficiencies [57 – 62].

We apply assembly strategies developed for flat surfaces to curved surfaces. Thiolate-linked anthracene phenylethynyl molecules were assembled on curved Au surfaces. Surface-enhanced Raman spectroscopy measurements revealed that the surface constraint drove unfavorable [4+4] photoreaction of anthracene rather than [4+2] reaction that is favored in solution on both flat and curved surfaces. Molecules on curved surfaces exhibit dramatically slower regioselective photoreaction kinetics and efficiencies than those on atomically flat surfaces due to larger intermolecular distances and variable orientations in the SAMs. The surface-dependent properties of such molecules can be used to probe the local chemical environment on curved surfaces.

Continuous efforts in the three-pronged approach involving assembly, measurements, and molecular design will enable the increasingly complex, precise, and multifunctional assemblies. These complex assemblies can serve as new test beds for gaining deeper insight into interactions and function of molecules. One of our long-term goals for this study are to design and to apply new functional materials controlled at the single-molecule level. By measuring single molecules and precise assemblies as test structures, we aim to elucidate the rules of working at the ultimate limits of miniaturization and to develop cooperative action inspired by biological systems.

Acknowledgments

We thank the Department of Energy (Grant No. SC 0005161) for support of the work described here.

REFERENCES

[1] Barth, J.V., Costantini, G., Kern, K. (2005) *Nature* **437**:671 – 679. doi: 10.1038/nature04166.

[2] Balzani, V., Credi, A., Venturi, M. (2008) *Molecular Devices and Machines: Concepts and Perspectives for the NanoWorld.* Wiley-VCH: Weinheim. doi: 10.1002/9783527621682.

[3] Weiss, P.S. (2001) *Nature* **413**:585 – 586. doi: 10.1038/35098175.

[4] Weiss, P.S. (2011) *Nature* **479**:187 – 188. doi: 10.1038/479187a.

[5] Weiss, P.S. (2008) *Acc. Chem. Res.* **41**:1772 – 1781. doi: 10.1021/ar8001443.

[6] Pathem, B.K., Claridge, S.A., Zheng, Y.B., Weiss, P.S. (2013) *Annu. Rev. Phys. Chem.* **64**:605. doi: 10.1146/annurev-physchem-040412-110045.

[7] Zheng, Y.B., Pathem, B.K., Hohman, J.N., Thomas, J.C., Kim, M., Weiss, P.S. (2013) *Adv. Mater.* **25**:302 – 312. doi: 10.1002/adma.201201532.

[8] Li, D., Paxton, W.F., Baughman, R.H., Huang, T.J., Stoddart, J.F., Weiss, P.S. (2009) *MRS Bull.* **34**:671 – 681. doi: 10.1557/mrs2009.179.

[9] Balzani, V., Gomez-Lopez, M., Stoddart, J.F. (1998) *Acc. Chem. Res.* **31**:405 – 414. doi: 10.1021/ar970340y.

[10] Balzani, V., Credi, A., Raymo, F.M., Stoddart, J.F. (2000) *Angew. Chem. Int. Ed.* **39**:3349 – 3391. doi: 10.1002/1521-3773(20001002)39:19<3348::AID-ANIE3348>3.0.CO;2-X.

[11] Claridge, S.A., Liao, W.-S., Thomas, J.C., Zhao, Y., Cao, H.H., Cheunkar, S., Serino, A.C., Andrews, A.M., Weiss, P.S. (2013) *Chem. Soc. Rev.* **42**:2725 – 2745. doi: 10.1039/C2CS35365B.

[12] Katsonis, N., Lubomska, M., Pollard, M.M., Feringa, B.L., Rudolf, P. (2007) *Prog. Surf. Sci.* **82**:407 – 434. doi: 10.1016/j.progsurf.2007.03.011.

[13] Klajn, R. (2010) *Pure Appl. Chem.* **82**:2247 – 2279. doi: 10.1351/PAC-CON-10-09-04.

[14] Morgenstern, K. (2011) *Prog. Surf. Sci.* **86**:115 – 161.
doi: 10.1016/j.progsurf.2011.05.002.

[15] Bumm, L.A., Arnold, J.J., Cygan, M.T., Dunbar, T.D., Burgin, T.P., Jones, L., Allara, D.L., Tour, J.M., Weiss, P.S. (1996) *Science* **271**:1705 – 1707.
doi: 10.1126/science.271.5256.1705.

[16] Donhauser, Z.J., Mantooth, B.A., Kelly, K.F., Bumm, L.A., Monnell, J.D., Stapleton, J.J., Price, D.W., Rawlett, A.M., Allara, D.L., Tour, J.M., Weiss, P.S. (2001) *Science* **292**:2303 – 2307.
doi: 10.1126/science.1060294.

[17] Ye, T., Kumar, A.S., Saha, S., Takami, T., Huang, T.J., Stoddart, J.F., Weiss, P.S. (2010) *ACS Nano* **4**:3697 – 3701.
doi: 10.1021/nn100545r.

[18] Claridge, S.A., Schwartz, J.J., Weiss, P.S. (2011) *ACS Nano* **5**:693 – 729.
doi: 10.1021/nn103298x.

[19] Han, P., Mantooth, B.A., Sykes, E.C.H., Donhauser, Z.J., Weiss, P.S. (2004) *J. Am. Chem. Soc.* **126**:10787 – 10793.
doi: 10.1021/ja049113z.

[20] Zheng, Y.B., Payton, J.L., Chung, C.-H., Liu, R., Cheunkar, S., Pathem, B.K., Yang, Y., Jensen, L., Weiss, P.S. (2011) *Nano Lett.* **11**:3447 – 3452.
doi: 10.1021/nl2019195.

[21] Pathem, B.K., Zheng, Y.B., Payton, J.L., Song, T.-B., Yu, B.-C., Tour, J.M., Yang, Y., Jensen, L., Weiss, P.S. (2012) *J. Phys. Chem. Lett.* **3**:2388 – 2394.
doi: 10.1021/jz300968m.

[22] Pathem, B.K., Zheng, Y.B., Morton, S., Petersen, M.Å., Zhao, Y., Chung, C.-H., Yang, Y., Jensen, L., Nielsen, M.B., Weiss, P.S. (2013) *Nano Lett.* **13**(2):337 – 343.
doi: 10.1021/nl304102n.

[23] Zheng, Y.B., Payton, J.L., Song, T.-B., Pathem, B.K., Zhao, Y., Ma, H., Yang, Y., Jensen, L., Jen, A.K.Y., Weiss, P.S (2012) *Nano Lett.* **12**:5362 – 5368.
doi: 10.1021/nl302750d.

[24] Hohman, J.N., Kim, M., Wadsworth, G.A., Bednar, H.R., Jiang, J., LeThai, M.A., Weiss, P.S. (2011) *Nano Lett.* **11**:5104 – 5110.
doi: 10.1021/nl202728j.

[25] Cygan, M.T., Dunbar, T.D., Arnold, J.J., Bumm, L.A., Shedlock, N.F., Burgin, T.P., Jones, L., Allara, D.L., Tour, J.M., Weiss, P.S. (1998) *J. Am. Chem. Soc.* **120**:2721 – 2732.
doi: 10.1021/ja973448h.

[26] Kim, M., Hohman, J.N., Cao, Y., Houk, K.N., Ma, H., Jen, A.K.Y., Weiss, P.S. (2011) *Science* **331**:1312 – 1315.
doi: 10.1126/science.1200830.

[27] Kumar, A.S., Ye, T., Takami, T., Yu, B.C., Flatt, A.K., Tour, J.M., Weiss, P.S. (2008) *Nano Lett.* **8**:1644 – 1648.
doi: 10.1021/nl080323+.

[28] Lewis, P.A., Inman, C.E., Yao, Y.X., Tour, J.M., Hutchison, J.E., Weiss, P.S. (2004) *J. Am. Chem. Soc.* **126**:12214 – 12215.
doi: 10.1021/ja038622i.

[29] Moore, A.M., Mantooth, B.A., Donhauser, Z.J., Yao, Y.X., Tour, J.M., Weiss, P.S. (2007) *J. Am. Chem. Soc.* **129**:10352 – 10353.
doi: 10.1021/ja0745153.

[30] Smith, R.K., Lewis, P.A., Weiss, P.S. (2004) *Prog. Surf. Sci.* **75**:1 – 68.
doi: 10.1016/j.progsurf.2003.12.001.

[31] Vaish, A., Shuster, M.J., Cheunkar, S., Singh, Y.S., Weiss, P.S., Andrews, A.M. (2010) *ACS Chem. Neurosci.* **1**:495 – 504.
doi: 10.1021/cn1000205.

[32] Alemani, M., Selvanathan, S., Ample, F., Peters, M.V., Rieder, K.-H., Moresco, F., Joachim, C., Hecht, S., Grill, L. (2008) *J. Phys. Chem. C* **112**:10509 – 10514.
doi: 10.1021/jp711134p.

[33] Love, J.C., Estroff, L.A., Kriebel, J.K., Nuzzo, R.G., Whitesides, G.M. (2005) *Chem. Rev.* **105**:1103 – 1169.
doi: 10.1021/cr0300789.

[34] Bumm, L.A., Arnold, J.J., Charles, L.F., Dunbar, T.D., Allara, D.L., Weiss, P.S. (1999) *J. Am. Chem. Soc.* **121**:8017 – 8021.
doi: 10.1021/ja9821571.

[35] Saavedra, H.M., Mullen, T.J., Zhang, P.P., Dewey, D.C., Claridge, S.A., Weiss, P.S. (2010) *Rep. Prog. Phys.* **73**:036501.
doi: 10.1088/0034-4885/73/3/036501.

[36] Haggui, M., Dridi, M., Plain, J., Marguet, S., Perez, H., Schatz, G.C., Wiederrecht, G.P., Gray, S.K., Bachelot, R. (2012) *ACS Nano* **6**:1299 – 1307.
doi: 10.1021/nn2040389.

[37] Xia, X.H., Yang, M.X., Wang, Y.C., Zheng, Y.Q., Li, Q.G., Chen, J.Y., Xia, Y.N. (2012) *ACS Nano* **6**:512 – 522.
doi: 10.1021/nn2038516.

[38] Wang, Y.F., Zeiri, O., Neyman, A., Stellacci, F., Weinstock, I.A. (2012) *ACS Nano* **6**:629 – 640.
doi: 10.1021/nn204078w.

[39] Gohler, B., Hamelbeck, V., Markus, T.Z., Kettner, M., Hanne, G.F., Vager, Z., Naaman, R., Zacharias, H. (2011) *Science* **331**:894 – 897.
doi: 10.1126/science.1199339.

[40] Shuster, M.J., Vaish, A., Gilbert, M.L., Martinez-Rivera, M., Nezarati, R.M., Weiss, P.S., Andrews, A.M. (2011) *J. Phys. Chem. C* **115**:24778 – 24787.
doi: 10.1021/jp207396m.

[41] Watari, M., McKendry, R.A., Vogtli, M.,Aeppli, G., Soh, Y.A., Shi, X.W., Xiong, G., Huang, X.J., Harder, R., Robinson, I.K. (2011) *Nature Mater.* **10**:862 – 866.
doi: 10.1038/nmat3124.

[42] Vaish, A., Liao, W.S., Shuster, M.J., Hinds, J.M., Weiss, P.S., Andrews, A.M. (2011) *Anal. Chem.* **83**:7451 – 7456.
doi: 10.1021/ac2016536.

[43] Vaish, A., Shuster, M.J., Cheunkar, S., Weiss, P.S., Andrews, A.M. (2011) *Small* **7**:1471 – 1479.
doi: 10.1002/smll.201100094.

[44] Hohman, J.N., Claridge, S.A., Kim, M., Weiss, P.S. (2010) *Mater. Sci. Eng. R* **70**:188 – 208.
doi: 10.1016/j.mser.2010.06.008.

[45] Donhauser, Z.J., Price, D.W., Tour, J.M., Weiss, P.S. (2003) *J. Am. Chem. Soc.* **125**:11462 – 11463.
doi: 10.1021/ja035036g.

[46] Mullen, T.J., Srinivasan, C., Hohman, J.N., Gillmor, S.D., Shuster, M.J., Horn, M.W., Andrews, A.M., Weiss, P.S. (2007) *Appl. Phys. Lett.* **90**:063114.
doi: 10.1063/1.2457525.

[47] Zheng, Y.B., Payton, J.L., Chung, C.H., Liu, R., Cheunkar, S., Pathem, B.K.,Yang, Y., Jensen, L., Weiss, P.S. (2011) *Nano Lett.* **11**:3447 – 3452.
doi: 10.1021/nl2019195.

[48] Hohman, J.N., Zhang, P., Morin, E.I., Han, P., Kim, M., Kurland, A.R., McClanahan, P.D., Balema, V.P., Weiss, P.S. (2009) *ACS Nano* **3**:527 – 536.
doi: 10.1021/nn800673d.

[49] Pathem, B.K., Kumar, A.S., Cao, Y., Zheng, Y.B., Corley, D.A., Ye, T., Aiello, V.J., Crespi, V., Houk, K.N., Tour, J.M., Weiss, P.S. 2013 (in preparation).

[50] Zheng, Y.B., Pathem, B.K., Payton, J.L. ,Bob, B., Kumar, A.S., Chung, C.H., Corley, D.A., Yang, Y., Jensen, L., Tour, J.M., Weiss, P.S. 2013 (in preparation).

[51] Cederquist, K.B., Keating, C.D. (2010) *Langmuir* **26**:18273 – 18280. doi: 10.1021/la1031703.

[52] Hill, H.D., Millstone, J.E., Banholzer, M.J., Mirkin, C.A. (2009) *ACS Nano* **3**:418 – 424. doi: 10.1021/nn800726e.

[53] Prigodich, A.E., Seferos, D.S., Massich, M.D., Giljohann, D.A., Lane, B.C., Mirkin, C.A. *ACS Nano* **3**:2147 – 2152. doi: 10.1021/nn9003814.

[54] Barth, B.M., Sharma, R., Alt\inolu, E.I., Morgan, T.T., Shanmugavelandy, S.S., Kaiser, J.M., McGovern, C., Matters, G. L., Smith, J.P., Kester, M., Adair, J.H. (2010) *ACS Nano* **4**:1279 – 1287.

[55] Weeraman, C., Yatawara, A.K., Bordenyuk, A.N., Benderskii, A.V. (2006) *J. Am. Chem. Soc.* **128**:14244 – 14245. doi: 10.1021/ja065756y.

[56] Templeton, A.C., Hostetler, M.J., Kraft, C.T., Murray, R.W. (1998) *J. Am. Chem. Soc.* **120**:1906 – 1911. doi: 10.1021/ja973863+.

[57] Wan, P., Jiang, Y., Wang, Y., Wang, Z., Zhang, X. (2008) *Chem. Commun.* 5710 – 5712. doi: 10.1039/b811729b.

[58] Sortino, S., Petralia, S., Conoci, S., Di Bella, S. (2004) *J. Mater. Chem.* **14**:811 – 813. doi: 10.1039/b314710j.

[59] Akiyama, H., Tamada, K., Nagasawa, J., Abe, K., Tamaki, T. (2003) *J. Phys. Chem. B* **107**:130 – 135. doi: 10.1021/jp026103g.

[60] Ito, M., Wei, T.X., Chen, P.-L., Akiyama, H., Matsumoto, M., Tamada, K., Yamamoto, Y. (2005) *J. Mater. Chem.* **15**:478 – 483. doi: 10.1039/b411121d.

[61] Wagner, S., Leyssner, F., Koerdel, C., Zarwell, S., Schmidt, R., Weinelt, M., Rueck-Braun, K., Wolf, M., Tegeder, P. (2009) *Phys. Chem. Chem. Phys.* **11**:6242 – 6248. doi: 10.1039/b823330f.

[62] Baisch, B., Raffa, D., Jung, U., Magnussen, O.M., Nicolas, C., Lacour, J., Kubitschke, J., Herges, R. (2008) *J. Am. Chem. Soc.* **131**:442 – 443. doi: 10.1021/ja807923f.

Beilstein Bozen Symposium on Molecular Engineering and Control
May 14th – 18th, 2012, Prien (Chiemsee), Germany

ORGANIC CHARGE TRANSFER SYSTEMS: THE NEXT STEP IN MOLECULAR ELECTRONICS?

MICHAEL HUTH

Physikalisches Institut, Goethe-Universität, Max-von-Laue-Str. 1,
60438 Frankfurt am Main, Germany

E-MAIL: michael.huth@physik.uni-frankfurt.de

Received: 6th December 2012 / Published: 13th December 2013

ABSTRACT

Since the 1950s a lot of effort has been devoted into the development of organic conductors. By a substantial degree this has been inspired by W. A. Little's theoretical analysis of London's idea that superconductivity might occur in organic macromolecules or polymers [1]. Over time this has resulted in the discovery of several organic metals and superconductors, albeit not with spectacularly high critical temperatures. The dominating material class in this respect are the organic charge transfer (CT) compounds which have provided an extended playground for studying the physics of low-dimensional, strongly correlated electron systems [2]. On the other hand, the practical application of organic materials relies on organic semiconductors rather than organic metals. Presently, polymers have found access to the semiconductor industry in capacitors and rechargeable batteries. Current research interests, which have been initiated already in the 1980s, are devoted to an application-oriented development for the growing market of organic electroluminescence devices, field effect transistors and solar cells. These devices are largely based on one-component organic materials or blends. Quite recently the research field of single-molecule electronics becomes increasingly popular.

There is a certain irony in the fact that by far the largest fraction of organic CT systems is semiconducting or insulating, whereas only a small fraction shows conducting properties or even becomes superconducting. As a consequence, research on this material class has chiefly been devoted to studying the properties of the small subset of conducting representatives of this material class. From the crystal structure point of view semiconducting or insulating CT compounds are mostly of the mixed-stack type, which means that donor and acceptor molecules are alternately arranged in one-dimensional chains. Via hybridization of the highest occupied molecular orbital (HOMO) of the donor with the lowest unoccupied molecular orbital (LUMO) of the acceptor a fully occupied binding molecular orbital results which causes a filled valence band in the solid and renders the material semiconducting or insulating. On the one hand, organic CT systems might become interesting alternatives to the one-component organic semiconductors. This is due to the fact that the additional Coulomb contribution to the binding energy in the CT compounds causes a somewhat larger electronic bandwidth as compared to the one-component molecular crystals which promotes an enhanced charge carrier mobility. More important, however, is the additional functionality which can be gained from the CT systems, e. g. a nonlinear conductivity behavior reminiscent of a Gunn diode or a thyristor.

This presentation gives a very brief introduction into some aspects of organic CT systems and will provide a short overview of the different electronic structures which can be observed in this material class. The major part will be devoted to introducing the so called neutral-ionic phase transition CT systems which are of the mixed-stack type and show a pressure- or temperature-driven transition between two different charge transfer states. This gives rise to a wealth of unconventional charge transport phenomena which might have a role to play in future organic electronics.

INTRODUCTION

Organic solids

Constituents of organic solids are molecules or their ions (molecular ions or radical ions). They form single crystalline, polycrystalline or glass-like structures. In the following the focus will be on crystalline materials. The most important energy scales relevant for the formation of the crystals and their ensuing electronic band structure in close proximity to the Fermi level are the electron affinity E_A and ionization energy E_I. Both are measured from the vacuum level E_{vac}. E_A denotes the energetic position of the lowest unoccupied molecular orbital (LUMO) and E_I the position of the highest occupied molecular orbital (HOMO) below E_{vac}. Figure 1 shows this is schematically for anthracene, a polycyclic aromatic hydrocarbon, which forms an organic semiconductor in its crystallized form.

123

Organic Charge Transfer Systems: the Next Step in Molecular Electronics?

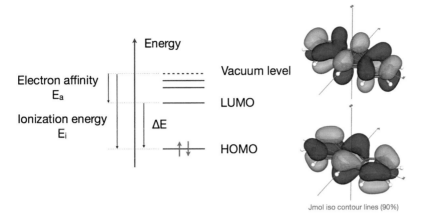

Figure 1. Schematic representation of relevant energy levels in donor or acceptor molecules. On the right, the highest occupied molecular orbital (HOMO) and lowest unoccupied molecular orbital (LUMO) of anthracene is exemplarily depicted as electron density iso-contour lines using Jmol.

With regard to their electronic properties the molecular energy scales defined by the HOMO and LUMO positions are essential for the ensuing electronic band structure in the crystal. Above all, Van-der-Waals interactions provide the most important contribution to the binding energy in molecular solids. As a consequence, intermolecular overlap and transfer integrals are small which results in rather narrow HOMO- and LUMO-derived bands with typical band widths in the range of 0.1 to 0.3 eV, whereas the energy gaps reflect the typical HOMO-LUMO energy differences of up to several eV.

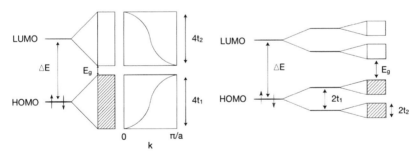

Figure 2. Schematic of the HOMO and LUMO level splitting towards weakly dispersing energy bands as a molecular crystal is assembled (left). The indicated band widths or band dispersions are highly exaggerated with typical band width of about 0.2 eV and band gaps in the eV range. On the right the band structure for a molecular crystal with dimerization is schematically indicated. This leads to additional band splittings.

Figure 2 depicts these aspects of the band structure in a qualitative fashion and also takes the consequences of dimerization of molecules into account which is of particular relevance in strong donor-acceptor crystals and radical ion salts to be introduced next. Many more details concerning the properties and applications of organic solids can be found in [3].

Organic charge transfer systems

In organic charge transfer systems (CTS), which result from the co-crystallization of donor (D) and acceptor (A) molecules, the crystal structures are determined by the strongly increased Coulomb and dipolar interactions. TTF-TCNQ (TTF: tetrathiafulvalene, TCNQ: tetracyanoquinodimethane), a one-dimensional metal that shows a Peierls transition below 54 K [4], is probably the most studied system within this material class. A low ionization energy of the donor accompanied by a large electron affinity of the acceptor favors the formation of a CTS. Spatial proximity for sufficient overlap between the respective donor and acceptor molecular orbitals of, ideally, similar symmetry is also necessary. This does already provide a first indication for the preferred spatial arrangement of the donor and acceptor molecules in the prevailing number of known CTS, namely the so-called mixed stack arrangement which will be discussed in somewhat more detail later. Figure 3 provides a schematic overview of the principal stacking arrangements to be found in the organic CTS.

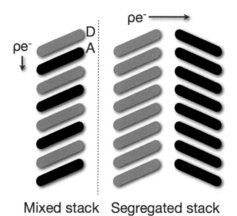

Mixed stack   Segregated stack

Figure 3. Schematic stacking arrangements of donor (D) and acceptor (A) molecules in a quasi one-dimensional crystal of an organic charge transfer system.

The discrimination between a weak and a strong CTS is represented by the nature of its ground state wave function which is a resonant state Φ_G between the neutral Φ_{DA} and ionized Φ_{D+A-} states

$$\Phi_G = a\Phi_{DA} + b\Phi_{D+A-} \tag{1}$$

For a>>b the CTS is of the weak type. These systems are not considered here. The actual degree of charge transfer between donor and acceptor is given as a fraction of an elemental charge δ per donor-acceptor pair.

As mentioned above, Coulomb and diploar interactions are important in CTS which has to be taken into account in the calculation of the binding energy E_B per D/A pair which is given by

$$E_B = (E_I - E_A)\delta - M\delta^2 \qquad (2)$$

The ionic contribution to the binding energy is reflected in the Madelung energy $E_M = M\delta^2$ (M: Madelung sum per D/A pair). E_M is the decisive factor in temperature- or pressure-driven (sudden or gradual) changes of the charge transfer degree δ which will be discussed in the next section on neutral-ionic phase transition systems.

In the case of a mixed-stack CTS a first qualitative description of the band structure can be obtained from assuming that inter-stack interactions can be neglected, so that the structure is that of a one-dimensional chain of D/A pairs. The resulting valence (VB) and conduction band (CB) are then derived from the overlap of the donor's HOMO and the acceptor's LUMO which are themselves formed by linear combinations of p_z atomic orbitals perpendicular to the molecular planes. Since the p orbitals have a node in the molecular plane, the quantum mixing of the D and A molecular orbitals depends on their relative lateral position perpendicular to the stacking axis, as well as on the inclination angle of the molecules towards this axis. As a consequence, the maximum dispersion of the bands need not be at the Brillouin zone center. Also, the weight of the donor's HOMO and the acceptor's LUMO in the VB and CB will vary in k-space, depending in parallel on the degree of charge transfer. This has been very transparently described by Katan and Koenig for the structurally simple organic charge transfer system TTF-2,5Cl2BQ (2,5Cl2BQ: 2,5-dichloro-p-benzoquinone) [5].

Besides the stacking type, the functional properties of organic CTS is strongly determined by the degree of charge transfer. This leads to the concept of ionicity. Full ionicity is obtained for $\delta = 1$. In the range $\delta_c < \delta < 1$ with δ_c 0.5 a CTS is said to be of the mixed-valence type. For $\delta < \delta_c$ a system is classified as neutral. This concept can be extended to the more general class of CTS of the type $(D^{\delta+})_m(A^{\delta-})_n$. Based on the stacking type and ionicity, Saito and Murata suggested a family tree of organic CTS which is reproduced in a different form in Fig. 4 [6]. Thus, the wealth of different ground states and associated functionalities of this material class becomes directly apparent. For further reading a recent comprehensive review by Saito and Yoshida is highly recommended [7]. In the next section the focus will be on the class of insulating mixed-valence systems with alternate stacking which can exhibit a so-called neutral-ionic phase transition which leads to a ferro- or antiferroelectric ground state.

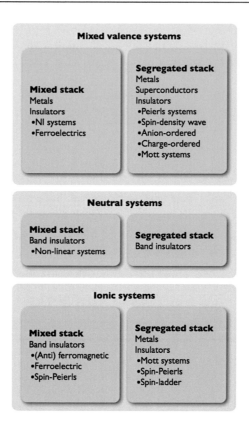

Figure 4. Hierarchical representation of organic charge transfer systems depending on the stacking type and degree of charge transfer.

NEUTRAL-IONIC PHASE TRANSITION SYSTEMS

In mixed-stack organic CTS a conversion from the neutral to the ionic phase (NIT: neutral-ionic transition) can occur, depending on the D and A type, as well as on the packing motif. The NIT can be driven by temperature, pressure and also by photo-irradiation [8, 9, 10]. Different continuous or discontinues types of NIT have been reported with varying degree of charge transfer between donor and acceptor. The transitions are often accompanied by D/A dimerization along the stacking axis. The mixed stack system TTF-CA (CA: p-chloranil) can be considered the prototypical compound of this material class. Under ambient pressure its NIT is of first order and shows a change of the charge-transfer degree from about 0.3 in the neutral phase to about 0.7 in the ionic phase. In electron spin resonance experiments a Curie-Weiss law is observed below the transition indicating the existence of unpaired spins. From these observations it can be concluded that the NIT in TTF-CA (and others) involves charge, spin and lattice degrees of freedom which renders the development of a microscopic under-standing of this material both, interesting and complex.

127

Organic Charge Transfer Systems: the Next Step in Molecular Electronics?

The phase transition at 81 K is accompanied by a loss of inversion symmetry resulting in a ferroelectric low-temperature phase of slightly deformed and dimerized D/A molecules [11]. In Fig. 5a the crystal structure of TTF-CA in the neutral phase is shown in conjunction with a qualitative indication of the structural changes associated with the NIT. Temperature-dependent anomalies in the dielectric constant and the electrical resistivity are schematically sketched in Figs. 5b and 5c.

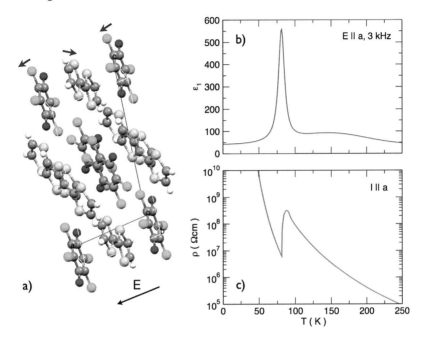

Figure 5. (a) Representation of the crystal structure of TTF-CA in the neutral phase. The arrows indicate the weak distortion in the molecular positions of about 1% in the ionic phase. **(b)** Qualitative plot of the temperature-dependent real part of the dielectric function ε_1 at moderate frequency, as indicated. The NI-transition is clearly visible as a sharp peak-like feature with ε_1 reaching values of several 100. **(c)** Qualitative plot of the temperature-dependent resistivity. At the NI-transition the resistivity shows a sharp anomaly.

The by far dominating amount of theoretical work trying to understand the underlying microscopic mechanisms that drive the NIT are based on modifications of the one-dimensional Hubbard model. TTF-CA can be considered essentially as one-dimensional system due to the π-orbital overlap which is, to a very large degree, restricted to the stacking direction. As a matter of fact, the narrow-band ideal of the Hubbard model is more closely realized by the electronic states in molecular crystals than in the d-electrons for which the Hubbard model is most often applied. A detailed account on the necessary modifications of the simple 1D-Hubbard model to take the D/A stacking and the Madelung energy (in a mean-field fashion) into account is given in [12]. This was later augmented by an electron –

molecular vibration coupling term which is also important for the NIT. The resulting modified Hubbard Hamiltonian has been shown to describe valence transitions and provides a good starting point for the description of the electron-phonon coupling in correlated systems. A recent example for the application of this Hubbard Hamiltonian approach on the symmetry crossover and excitation thresholds in the NIT can be found in [13]. The model-immanent limitation to treating the inherently long-ranged Madelung energy in a mean-field fashion represents one drawback in the Hubbard approach to the NIT. In fact, an *ab-initio* treatment of the problem without any prior assumptions of the relative strengths of the electronic interactions would also be desirable. It has been only rather recently that first-principles density functional theory calculations have become possible for organic CTS. Recent work in this regard has studied TTF-CA and found, e.g., that within the neutral phase just above the NIT the indirect band gap between the VB and CB tends to close as a consequence of thermal contraction of the lattice [14]. Interestingly, with the gap closing an increased charge transfer necessarily arises, so that one may argue that the NIT prevents the system from developing a metallic state.

At this stage we leave these rather general deliberations and focus on two properties of the NIT-systems which hold some promise with regard to applications: (i) the ferroelectric ground state and (ii) the electrical properties associated with domain formation in the neutral and ionic phase.

Ferroelectricity in TTF-CA

Switchable ferroelectric materials are very interesting for a range of electronic applications, such as in storage devices and sensor development. In this regard organic CTS of the NIT type hold some promise. At the same time, this material class also clearly reflects the complexities which arise in the proper definition and microscopic treatment of ferroelectricity. Before briefly discussing the ferroelectric properties of TTF-CA, a few words are in order to clarify the concept of ferroelectricity in general.

Macroscopic electric polarization P can in most instances not been simply defined via the dipole of a unit cell of a lattice. It is only in the limit of purely ionic, localized charges without a continuous electrical charge distribution between the lattice sites that such a definition holds. As reviewed by Resta [15] a proper definition takes both, the ionic and charge distribution components into account

$$P = \frac{1}{V}\left[-e\sum_j Zr_j + \int d^3r\, r\rho(r) \right] \tag{3}$$

where the sum describes the ionic contribution given by the ions on lattice sites indexed with j and $\rho(\mathbf{r})$ is the charge density of a sample with volume V. Ultimately, the charge density is obtained from the quantum mechanical wave function, so that polarization must be considered as a quantum mechanical phenomenon. Due to the continuity equation for the charge,

129

Organic Charge Transfer Systems: the Next Step in Molecular Electronics?

the macroscopic polarization is associated with a macroscopic (displacement) current density which in turn is governed by the phase of the wave function. From this viewpoint it becomes apparent that the concept of a geometrical Berry phase is a suitable starting point for a microscopic understanding of ferroelectricity.

When considering the two classes of orientational "order-disorder"-like transitions of permanent dipoles and "displacive"-type transitions, organic ferroelectrics belong to the latter type. Moreover, the rather strong hybridization between the donor and acceptor molecular orbitals lends special importance to the electronic charge density contribution to the polarization. This is particularly true for TTF-CA, as has been recently shown by Kabayashi and collaborators [16]. They found the ferroelectricity of TTF-CA to be dominated by the D/A charge transfer and only to a small degree by the contribution of localized ionic charge dipole moments. In low-stress single crystals of TTF-CA they measured the dielectric constant at several hundred kHz along the stacking direction and found a clear Curie-Weiss behavior of the type $\varepsilon_r = C/(T - \theta)$ above the NIT temperature with a large Curie-Weiss constant C. This is indicative of a large polarization in the ferroelectric phase. However, the observation of a clear P-E-hysteresis, a fingerprint of a ferroelectric material, is hindered in TTF-CA due to a large non-linear conductance contribution (see also next subsection) causing leakage currents already at a moderate electric field strength of 10 kV/cm. The authors were nevertheless able to measure hysteresis loops at temperatures below about 60 K. By carefully analyzing the data they found clear evidence that the electronic contribution to polarization exceeds the ionic part by a factor of about 20. This labels the ferroelectric state in TTF-CA as an emergent phenomenon resulting from electronic correlation effects. Also, the associated weak lattice distortion in this type of ferroelectric leads to high resonance frequencies for the polarization response. NIT-systems with ferroelectric ground state and transition temperatures above room temperature are therefore very promising for high-frequency applications, such as in Fe-RAM devices operating at high switching speeds.

Charge-transport phenomena in TTF-CA

In TTF-CA single crystals the conductivity σ in the ionic phase, as well as in the neutral phase, is dominated by strong non-linear effects already at rather small electric fields E. This is schematically shown in Fig. 6 for the ionic phase at a temperature close to the NIT. An ohmic behavior at low fields is followed by a quasi-linear $\sigma(E)$ behavior which shows a sudden switching to an "on"-state with three orders of magnitude larger conductivity at fields above about 10 kV/cm, depending on temperature. The transition to the less conducting "off"-state occurs at a slightly smaller field as the field is reduced again, i.e. the switching is hysteretic. Similar switching effects can be observed in the neutral phase. The rather complex interplay of charge, spin and lattice degrees of freedom is ultimately responsible for these effects. In order to understand some important aspects of the type of charge-carrying excitations in this material it is useful to briefly discuss a theoretical model of the NIT-systems introduced by Nagaosa [17].

We neglect possible inter-stack interactions and investigate a one-dimensional model based on the following Hamiltonian [17]

$$H = -\sum_{l,s} t_{l,l+1} \left[c_{l,s}^+ c_{l+1,s} + c_{l+1,s}^+ c_{l,s} \right]$$

$$+ \frac{\Delta_0}{2} \sum_l (-1)^l n_l + U \sum_l n_{l,\uparrow} n_{l,\downarrow}$$

$$(4)$$

$$+ V \left[\sum_{l \, even} n_l (n_{l+1} - 2) + \sum_{l \, odd} (n_l - 2) n_{l+1} \right]$$

$$+ \frac{2}{S} \sum_l (u_l - u_{l+1})^2$$

The first term describes charge carrier hopping along the stack. The transfer is assumed to have the form $t_{l,l+1} = T + (u_l - u_{l+1})$ with a transfer integral T and a linear electron-lattice coupling that tends to enhance or dampen the transfer depending on the sign of the relative lattice displacements on neighboring sites. The electron lattice coupling strength is set to S. $n_l = n_{l,\uparrow} + n_{l,\downarrow}$ and $n_{l,s} = c_{l,s} + c_{l,s}$ denote the fermionic number operators. The second term reflects the difference in ionization potential and electron affinity on the D sites (odd) and A sites (even), respectively, with $\Delta_0 = E_I - E_A + U$. U denotes the on-site Coulomb repulsion. The Madelung energy is taken into account on a nearest-neighbor level via the third term in the Hamiltonian. Finally, the last term gives the direct energy contribution to the electron system by way of the electron-phonon coupling. Since the lattice dimerization in the ionic phase amounts to only about 1% this last term is rather small. However, the transfer modulation due to the lattice dimerization is much more relevant being essentially a phase-factor effect in the hybridization of neighboring sites. Without going into more detail, which can be obtained from Nagaosa's excellent paper [17], some important statements can be made from this Hamiltonian in the case of half filling, i.e. the number of electrons corresponds to the number of lattice sites N.

(a) Neglecting the transfer and lattice term, some simple algebra on the then classical Hamiltonian shows that the energies per site in the ionic and neutral phase are

$$E_i/N = (U - \Delta_0)/2$$

$$E_n/N = -V$$

The phase transition occurs for $E_i = E_n$ from which the condition $(E_I - E_A)/2 = V$ is readily obtained, with the neutral phase being more stable for $(E_I - E_A)/2 > V$. This states that under lattice contraction the increase of the Madelung energy favors the ionic state.

131

Organic Charge Transfer Systems: the Next Step in Molecular Electronics?

(b) If the transfer term is taken into account, the ionic phase exhibits a gapless spin excitation (magnon) as is also the case in the one-dimensional Heisenberg model.

(c) The NIT is first order in this model.

(d) In the neutral phase the energy for a magnetic excitation rises steeply. The charge-transfer energy, on the other hand, is minimal at the phase boundary.

From these observations Nagaosa concluded that the lowest lying charge excitation in TTF-CA at the NIT are neutral-ionic phase domain boundaries with an energy about four times smaller than the energy necessary to split a charge-transfer excitation into a separated electron-hole pair. Due to the transfer term, the domain wall excitations acquire a dispersion. For further details the reader is referred to [17].

Based on Nagaosa's analysis Tokura and collaborators interpreted their observations of non-linear effects in the field-dependent conductivity of TTF-CA single crystals in a domain-wall picture [18], as is also schematically indicated in Fig. 6 using a phenomenological model for the current-voltage characteristics introduced by Iwasa *et al.* [19]. In order to account for the non-linearity they assumed a positive feedback effect caused by the electric field in the excitation process of the domain walls.

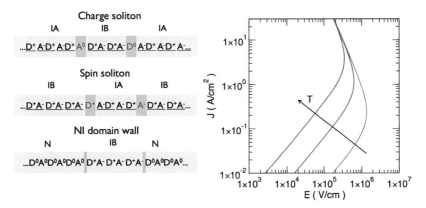

Figure 6. Schematic representation of domain-like excitations of a NI-transition systems with ionic ground state according to [18]. The dynamics of these excitations leads to a strongly non-linear current-voltage characteristics, as schematically indicated on the right.

From the point of view of using the ferroelectricity in NIT-systems for applications in which field-induced polarization reversal is necessary, the presence of low-lying charge excitations based on the domain-wall mechanism is unfavorable. However, in other NIT-systems, such as TTF-QBrCl3, these conductance contributions are less pronounced [16]. Furthermore, applications can be envisioned for which the field-induced conductivity switching itself can provide a useful thyristor-like functionality. In any case, the availability of thin films is mandatory, if NIT systems are to be used in organic electronics. This aspect is discussed next.

THIN FILMS OF NIT SYSTEMS

Thin film deposition of organic CTS has so far not received a lot of attention. This may to some degree be attributed to the fact that crystalline organic thin film growth poses some additional complications when compared to the more classical (inorganic) semiconductor and metal epitaxy. Nevertheless, thin film growth studies of TTF-TCNQ have been reported quite early after the discovery of this one-dimensional organic metal [20] and have been also actively pursued in some recent works [21 – 24]. With a view to the promising functionalities provided by this material class, e. g. in photovoltaics [25], this is likely to change soon. Some of our recent work has also been dealing with other D/A systems, such as (BEDT-TTF)TCNQ (BEDT-TTF: bisethylene-TTF) [26] or also new organic CTS [27]. In the following some very recent and so far unpublished results on the thin film growth of TTF-CA are presented [28].

The films have been prepared from mixed powders of TTF and CA in stoichiometric proportions by sublimation from one effusion cell at cell temperatures between 95 °C and 105 °C. Different substrate materials with different crystallographic orientations have been used. The substrates have been held at room temperature. The film growth was performed under ultra-high vacuum conditions at a base pressure of about 10^{-7} mbar. Figure 7 shows an X-ray diffractogram of a TTF-CA thin film grown on NaCl (100) (annealed). The preferential growth orientation is (211) which corresponds to the D/A molecules lying flat on the substrate surface. As evident from atomic force microscopy scans, also shown in Fig. 7, the growth proceeds by island formation. Interestingly, a partial wetting layer of 3 to 4 nm height forms between the islands. The composition of this layer is still under investigation. First temperature-dependent resistivity measurements have been performed on films grown on SiO$_2$/Si(100), as also shown in Fig. 7. Apparently, the NIT is not suppressed in the layers and occurs at about 81 K (onset) as in single crystals. Further work is in progress to optimize the growth and also study the dielectric properties of the TTF-CA thin films.

133

Organic Charge Transfer Systems: the Next Step in Molecular Electronics?

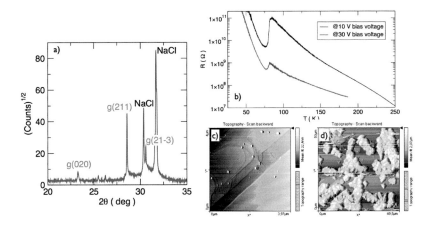

Figure 7. (a) X-ray diffractogram of TTF-CA thin film on annealed NaCl (100). The reflexion of the NI-phase are indicated. A (211) growth preference is observed which corresponds to a flat-on arrangement of the donor and acceptor molecules of the two non-equivalent D/A stacks (see Fig. 5a for reference). **(b)** Temperature-dependent resistivity of TTF-CA thin film grown on amorphous $SiO_2/Si(100)$. The data was taken under constant bias voltage conditions, as indicated. The NI-transition is clearly visible. The phase transition signature is suppressed at higher bias voltage. **(c)** Atomic force microscopy (AFM) image of the annealed NaCl(100) substrate used for the film growth. **(d)** AFM image of the island growth of TTF-CA layers on NaCl(100). The data was taken in non-contact mode.

CONCLUSION

NIT-systems represent a highly interesting material class from the group of mixed-stack organic charge transfer systems. Until recently the number of representatives has been rather small but a systematic search has resulted in a significant number of new neutral-ionic phase transition systems [29]. In the future, the thin film route will provide a valuable extension of this research with regard to tuning the NIT by film-specific factors, such as biaxial-strain formation via clamping, and the realization of device structures, such as field-effect transistors. In this sense, NIT-systems represent exciting materials for studying electronic correlation effects under partially competing couplings between the charge, spin and lattice degrees of freedom. At the same time, their application potential is significant and could be appreciably enhanced, if it turned out to be possible to shift the NIT to above room temperature.

ACKNOWLEDGMENTS

The author acknowledges financial support by the Beilstein-Institut, Frankfurt am Main, within the research collaboration NanoBiC. Financial support by the DFG through the transregional collaborative research center TR-SFB 49 is also gratefully acknowledged.

REFERENCES

[1] Little, W.A. (1964) Possibility of Synthesizing an Organic Superconductor. *Phys. Rev.* **134**:A1416.
doi: 10.1103/PhysRev.134.A1416.

[2] see, e.g., Sato, G., Yoshida, Y. (2007) Development of Conductive Organic Molecular Assemblies. Organic Metals, Superconductors, and Exotic Functional Materials. *Bull. Chem. Soc. Jpn.* **80**:1.
doi: 10.1246/bcsj.80.1.

[3] see, e.g., Schwoerer, M., Wolf, H.-Ch. (2006) Organic Molecular Solids (Physics Textbook), 1. edition, Wiley-VCH.

[4] see, e.g., Khanna, S.K., Pouget, J.P., Comes, R., Garito, A.F., Heeger, A.J. (1977) X-ray studies of 2kF and 4kF anomalies in tetrathiafulvalene-tetracyanoquinodimethane (TTF-TCNQ). *Phys. Rev. B* **16**:1468.
doi: 10.1103/PhysRevB.16.1468.

[5] Katan, C., Koenig, C. (1999) Charge-transfer variation caused by symmetry breaking in a mixed-stack organic compound: TTF-2,5Cl2BQ. *J. Phys. Cond. Matt.* **11**:4163.
doi: 10.1088/0953-8984/11/21/305.

[6] Saito, G., Murata, T. (2010) Mixed valency in organic charge transfer complexes. *Phil. Trans. R. Soc. A* **366**:139.
doi: 10.1098/rsta.2007.2146.

[7] Saito, G., Yoshida, Y. (2007) Development of Conductive Organic Molecular Assemblies: Organic Metals, Superconductors, and Exotic Functional Materials. *Bull. Chem. Soc. Jpn.* **80**:1.
doi: 10.1246/bcsj.80.1.

[8] Mayerle, J.J., Torrance, J.B., Crowley, J.I. (1979) Mixed-stack complexes of tetrathiafulvalene. The structures of the charge-transfer complexes of TTF with chloranil and fluoranil. *Acta Crystallogr. B* **35**:2988.
doi: 10.1107/S0567740879011110.

[9] Torrance, J.B., Vasquez, J.E., Mayerle, J.J., Lee, V.Y. (1981) Discovery of a Neutral-to-Ionic Phase Transition in Organic Materials. *Phys. Rev. Lett.* **46**:253.
doi: 10.1103/PhysRevLett.46.253.

[10] Koshihara, S.-Y., Takahashi, Y., Sakai, H., Tokura, Y., Luty, T. (1999) Photoinduced Cooperative Charge Transfer in Low-Dimensional Organic Crystals. *J. Phys. Chem. B* **103**:2592.
doi: 10.1021/jp984172i.

135

Organic Charge Transfer Systems: the Next Step in Molecular Electronics?

[11] LeCointe, M., Lemée-Cailleau, M.H., Cailleau, H., Toudic, B., Toupet, L., Heger, G., Moussa, F., Schweiss, P., Kraft, K.H., Karl, N. (1995) Symmetry breaking and structural changes at the neutral-to-ionic transition in tetrathiafulvalene-p-chloranil. *Phys. Rev. B* **51**:3374.
doi: 10.1103/PhysRevB.51.3374.

[12] Strebel, P.J., Soos, Z.G. (1970) Theory of Charge Transfer in Aromatic Donor-Acceptor Crystals. *J. Chem. Phys.* **53**:4077.
doi: 10.1063/1.1673881.

[13] Anusooya-Pati, Y., Soos, Z.G., Painelli, A. (2001) Symmetry crossover and excitation thresholds at the neutral-ionic transition of the modified Hubbard model. *Phys. Rev. B* **63**:205118.
doi: 10.1103/PhysRevB.63.205118.

[14] Oison, V., Katan, C., Rabiller, P., Souhassou, M., Koenig, C. (2003) Neutral-ionic phase transition: A thorough ab initio study of TTF-CA. *Phys. Rev. B* **67**:035120.
doi: 10.1103/PhysRevB.67.035120.

[15] Resta, R. (1994) Macroscopic polarization in crystalline dielectrics: the geometric phase approach. *Rev. Mod. Phys.* **66**:899.
doi: 10.1103/RevModPhys.66.899.

[16] Kobayashi, K., Horiuchi, S., Kumai, R., Kagawa, F., Murakami, Y., Tokura, Y. (2012) Electronic Ferroelectricity in a Molecular Crystal with Large Polarization Directing Antiparallel to Ionic Displacement. *Phys. Rev. Lett.* **108**:237601.
doi: 10.1103/PhysRevLett.108.237601.

[17] Nagaosa, N. (1986) Domain wall picture of the neutral-ionic transition in TTF-Chloranil. *Solid State Commun.* **57**:179.
doi: 10.1016/0038-1098(86)90134-1.

[18] Tokura, Y., Okamoto, H., Koda, T., Mitani, T., Saito, G. (1986) Domain wall dynamics in mixed-stack charge-transfer crystal. *Physica B* **143**:527.
doi: 10.1016/0378-4363(86)90187-7.

[19] Iwasa, Y., Koda, T., Koshihara, S., Tokura, Y., Iwasawa, N., Saito, G. (1989) Intrinsic negative-resistance effect in mixed-stack charge-transfer crystals. *Phys. Rev. B* **39**:10441.
doi: 10.1103/PhysRevB.39.10441.

[20] Chen, T.H., Schechtman, B.H. (1975) Preparation and properties of polycrystalline films of TTF-TCNQ. *Thin Solid Films* **30**:173.
doi: 10.1016/0040-6090(75)90319-3.

[21]   Fraxedas, J., Molas, S., Figueras, A., Jimenes, I., Gago, R., Auban-Senzier, P., Goffman, M. (2002) Thin films of molecular metals: TTF-TCNQ. *J. Solid State Chem.* **168**:384.
doi: 10.1006/jssc.2002.9685.

[22]   Solovyeva, V., Huth, M. (2011) Defect-induced shift of the Peierls transition in TTF-TCNQ thin films. *Synth. Met.* **161**:976.
doi: 10.1016/j.synthmet.2011.03.003.

[23]   Solovyeva, V., Cmyrev, A., Sachser, R., Reith, H., Huth, M. (2011) Influence of irradiation-induced disorder on the Peierls transition in TTF-TCNQ microdomains. *J. Phys. D: Appl. Phys.* **44**:385301.
doi: 10.1088/0022-3727/44/38/385301.

[24]   Sarkar, I., Laux, M., Demokritova, J., Ruffing, A., Mathias, S., Wei, J., Solovyeva, V., Rudloff, M., Naghavi, S.S., Felser, C., Huth, M., Aeschlimann, M. (2010) Evaporation temperature-tuned physical vapor deposition growth engineering of one dimensional non-Fermi liquid TTF-TCNQ thin films. *Appl. Phys. Lett.* **97**:111906.
doi: 10.1063/1.3489098.

[25]   Yuan, Y., Reece, T.J., Sharma, P., Poddar, S., Ducharme, S., Gruverman, A., Yang, Y., Huang, J. (2011) Efficiency enhancement in organic solar cells with ferroelectric polymers. *Nature Materials* **10**:296.
doi: 10.1038/nmat2951.

[26]   Solovyeva, V., Keller, K., Huth, M. (2009) Organic charge transfer phase formation in thin films of the BEDT-TTF/TCNQ donor-acceptor system. *Thin Solid Films* **517**:6671.
doi: 10.1016/j.tsf.2009.05.006.

[27]   Medjanik, K., Perkert, S., Naghavi, S., Rudloff, M., Solovyeva, V., Chercka, D., Huth, M., Nepijko, S. A., Methfessel, T., Felser, C., Baumgarten, M., Müllen, K., Elmers, H.-J., Schönhense, G. (2010) A new charge-transfer complex in UHV co-deposited tetramethoxypyrene and tetracyanoquinodimethane. *Phys. Rev. B* **82**:245419.
doi: 10.1103/PhysRevB.82.245419.

[28]   Keller, Lukas (2012) Dünne Schichten der organischen Ladungstransferverbindung TTF-Chloranil: Bachelor thesis, Goethe University, Frankfurt am Main.

[29]   Horiuchi, S., Kumai, R., Okimoto, Y., Tokura, Y. (2006) Chemical approach to neutral-ionic valence instability, quantum phase transition, and relaxor ferroelectricity in organic charge-transfer complexes. *Chem. Phys.* **325**:78.
doi: 10.1016/j.chemphys.2005.09.025.

Beilstein Bozen Symposium on Molecular Engineering and Control
May 14th – 18th, 2012, Prien (Chiemsee), Germany

137

Simulating "Soft" Electronic Devices

Timothy Clark[1,2]*, Marcus Halik[2,3], Matthias Hennemann[1], and Christof M. Jäger[1,2]

[1]Computer-Chemie-Centrum and Interdisciplinary Center for Molecular Materials, Department Chemie und Pharmazie, Friedrich-Alexander-Universität Erlangen-Nürnberg, Nägelsbachstrasse 25, 91052 Erlangen, Germany.

[2]Excellence Cluster "Engineering of Advanced Materials", Friedrich-Alexander-Universität Erlangen-Nürnberg, Nägelsbachstrasse 49b, 91052 Erlangen, Germany.

[3]Organic Materials & Devices, Institute of Polymer Materials, Department of Materials Science, Friedrich-Alexander-Universität Erlangen-Nürnberg, Martensstrasse 7, 91058 Erlangen, Germany.

E-Mail: *tim.clark@chemie.uni-erlangen.de

Received: 14th August 2013 / Published: 13th December 2013

Abstract

A combination of classical molecular dynamics simulations and very large scale semiempirical molecular orbital calculations has been used to simulate field-effect transistors in which both the dielectric layer and the semiconductor are incorporated in a self-assembled monolayer of suitable functionalized alkylphosphonic acids. In such simulations, both the dynamics of the flexible organic molecules and the electronic properties of the molecular aggregates must be taken into account. First steps towards realistic simulations of such devices are described.

Introduction

Modeling and simulation of molecules and molecular aggregates requires that two aspects of the modeling task be treated adequately; the Hamiltonian(s) used to describe the system and the conformational sampling. Whereas conformational sampling only plays a small role in many cases in which small molecules are modeled, it becomes the dominant aspect of modeling macromolecules, so that the quality of the Hamiltonian is dictated by the need for extensive sampling. This is generally not a problem as modern force fields provide adequate energy hypersurfaces at very reasonable computational cost, so that in particular biopolymers can be simulated for several microseconds on specialized hardware [1]. Such classical molecular-dynamics (MD) techniques have become essential in biological and medicinal research, but are of limited use in simulating organic electronics devices because they treat the electrons implicitly in what is essentially a coarse-grained ansatz designed to mimic quantum mechanics. It is therefore necessary either to perform direct quantum mechanical MD (e. g. Car-Parinnello density-functional theory (DFT) based MD) or to use a multiscale treatment in which the MD (i. e. the conformational sampling) is performed classically and the geometries of "snapshots" from the classical MD-simulation used for hundreds or thousands of single-point quantum mechanical calculations to determine the electronic properties. Because the classical simulation typically treats tens of thousands of atoms, the "snapshots" are very large for quantum mechanical treatments. This means that either only the most interesting components of the snapshot (e. g. the chromophores [2]) are treated quantum mechanically in a hybrid quantum mechanics/molecular mechanics (QM/MM [3]) approach or that very large scale quantum mechanical calculations must be used. This limits the choice of technique available. Linear scaling *ab initio* [4], or DFT [5] calculations are able to treat thousands of atoms, but not yet to perform calculations on thousands of snapshots, each consisting of tens of thousands of atoms. This means that "cheaper" quantum mechanical methods such as semiempirical molecular orbital (MO) theory [6] or tight-binding-based DFT [7] must be used. Because the self-interaction error limits the usefulness of DFT-based techniques for applications in which electrons are transferred from one moiety or region to another, [8] semiempirical MO-theory is our method of choice.

Linear scaling versions of semiempirical MO theory are usually based on the divide-and-conquer (D&C) [9,10] or localized molecular orbital (LMO) [11] techniques. However, because these both use local approximations, they are not suitable for electronics applications in which conjugation over large distances, areas and volumes is important. This is simply because the long-range interactions that determine the characteristics of extensively conjugated systems are explicitly excluded from local approximations because they can be neglected in non-conjugated systems. This means that full semiempirical self-consistent field (SCF) calculations must be performed. However, these scale with N^2 to N^3, depending on the implementation, so that very large-scale calculations rapidly become prohibitively expensive. In this article, we describe our simulations of field-effect transistors (FETs) constructed using self-assembled monolayers (SAMs) of organic molecules as an example of the challenges of such applications and the strategies that can be used to overcome them.

THE SAMFETs

The self-assembled monolayer field-effect transistors (SAMFETs) consist of gold source and drain electrodes, an oxidized aluminum drain electrode and a multifunctional SAM that provides both the insulating dielectric and the semiconductor layers [12]. Figure 1 shows a schematic view of a typical SAMFET.

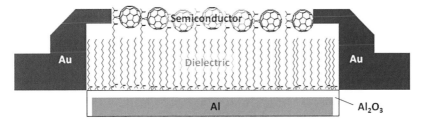

Figure 1. Schematic and idealized view of a SAMFET with an n-type semiconductor layer consisting of fullerene moieties.

The first important point illustrated in figure 1 is that the schematic view of the SAM is idealized in the extreme. The alkane chains do not all align in an orderly fashion perpendicular to the Al_2O_3 surface; the SAM is a dynamic and only moderately ordered structure. Similarly, the orderly row of fullerenes that act as the semiconductor is an ideal situation that is unlikely to be realized in a real device. An important point is that the SAM consists of a fullerene-functionalized molecule and one with a pure alkane tail. It was found experimentally that mixtures of such compounds gave functioning SAMFETs, whereas a SAM consisting of only fullerene-substituted molecules give only very poor transistor-like characteristics with very high gate currents. The interpretation shown in figure 1 that the unsubstituted molecules form a cushioning layer below the fullerenes is simply a rationalization of these experimental results. Figure 1 shows a fullerene-substituted SAM, which gives an n-type semiconductor layer. As an alternative to the fullerenes, oligothiophenes can be used to give a p-type semiconductor layer [13].

Rational design of SAMFETs of this type requires detailed knowledge at the atomistic level of the structure and electronic properties of the SAMs. This information is difficult to obtain experimentally, so that simulations provide a valuable source of information. This is true of purely classical simulations of the dynamic structure of the SAMs but even more so if electronic properties can also be calculated. This combination of structural dynamics and electronic properties represents a computational challenge that is only now becoming accessible because of the constant improvement in both hard- and software. In the following, we describe some aspects of this challenge and the techniques developed so far to answer it.

STRUCTURE OF THE SAMs: CLASSICAL MOLECULAR DYNAMICS

The MD simulations of SAMs used the 0001 surface of Al_2O_3 to represent the oxidized surface of the aluminum gate electrode. The 52×38 nm Al_2O_3 slab (five layers) was first equilibrated in the gas phase and then held fixed during the SAM simulations. One potential difficulty in such simulations is the treatment of the bonding between the Al_2O_3 and the organic molecules that make up the SAM. The alternatives are to define classical "covalent" bond potentials (in reality quadratic potentials according to Hooke's law) that cannot be broken or to treat the interaction as purely electrostatic. The latter alternative has the advantage that the organic molecules can dissociate from the surface or rearrange, so that for instance the surface coverage can adjust to an equilibrium value. Purely electrostatic binding was therefore used in the simulations. The Generalized Amber Force Field (GAFF [14]) was used for the organic molecules.

The SAMs consisting of anions **1 – 4** were constructed as regular arrays on the Al_2O_3 surface and allowed to equilibrate. Some molecules dissociated from the surface but stable SAMs were obtained in all simulations.

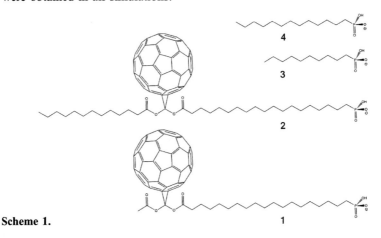

Scheme 1.

The simulations confirmed the main thrust of the interpretation shown schematically in figure 1 but also revealed details not accessible otherwise. Figure 2 shows a snapshot from an equilibrated simulation using a SAM consisting of pure **1**. The surface coverage is fairly low (0.9 molecules nm^{-2}) and the fullerenes do not form a single layer, but are rather distributed at different levels above the Al_2O_3 surface. More importantly, in this simulation individual fullerenes already contact the Al_2O_3 surface during the warm up phase (starting from an ordered vertical SAM). Further surface contacts are made within 20 ns simulation time. Snapshot showing one such occurrence is shown in figure 2.

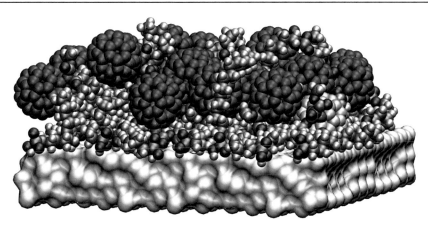

Figure 2. Snapshot taken from a simulation of a SAM consisting of pure **1** after 50 ns. The fullerene moiety that contacts the Al_2O_3 surface is colored yellow.

The simulation using a mixed SAM (25% **1**, 75% **3**) behaves differently. The initially perpendicular SAM adopts an average angle to the Al_2O_3 surface of approximately 58 ° after approximately 10 ns and remains at this angle throughout the remaining simulation time. More importantly, the fullerene moieties remain separated from the Al_2O_3 surface by a closed and consistent layer of alkyl chains, as shown in a representative snapshot (Fig. 3).

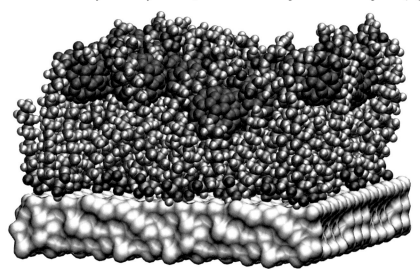

Figure 3. Snapshot taken from a simulation of a SAM consisting of 25% **1** and 75% **3** after 50 ns. The fullerene moieties form a defined layer.

This behavior is completely consistent with the experimental observations. A simple optical analysis of the simulation shows that the alkyl chains remain as a moderately ordered continuous layer below the fullerenes, which form a second fairly regular layer above them. This is the configuration required for the SAM to act as an FET. However, the classical simulations do not allow us to analyze the electronic properties of the SAM.

VERY LARGE SCALE SEMIEMPIRICAL MO CALCULATIONS

Modern NDDO-based semiempirical MO-calculations are almost exclusively derived from the original MNDO technique. [6, 15, 16] As discussed above, linear scaling techniques that rely on local approximations [9, 10, 11] are not suitable for extensively conjugated systems, so that software is needed that performs the full NDDO-based SCF calculation on many thousands of atoms. This and the current hardware emphasis on multi-core architectures mean that a highly parallel program is necessary. Unfortunately, little emphasis has been placed on parallel semiempirical MO codes, probably because the major computational task is the diagonalization of the Fock matrix. Matrix diagonalizations perform relatively poorly on highly parallel architectures.

We therefore conceived the EMPIRE program, [17] to perform very large (designed for up to 50,000 atoms, tested up to 100,000) on massively parallel (designed and tested up to 1,024 processors) compute clusters. Because the diagonalization of the Fock matrix represents the major bottleneck in the parallel calculation, the SCF iterations were designed to proceed with as few diagonalizations as possible. The details of the implementation have been reported elsewhere but the major design considerations were as follows:

- Memory requirements increase with N^2, where N is the number of atomic orbitals. Memory requirements were therefore reduced to a minimum. Ideally, only four $N \times N$ matrices are used.

- Memory access is slow relative to floating-point calculations, so that most quantities required during the calculation (e. g. two-electron integrals, the one-electron matrix) are calculated on the fly. The one-electron matrix can optionally be stored and reused in every SCF cycle.

- A hybrid parallelization strategy (openMP within a node, MPI between nodes) was used to optimize performance.

- The calculations rely heavily on Intel's optimized MKL parallel library for the majority of calculations within one node.

- The important matrices are distributed across the compute nodes in stripes to minimize communication.

An important component of the EMPIRE program is an accurate initial guess (the starting set of molecular orbitals for the SCF iterations) that allows it to converge for most systems very quickly without need for convergence accelerators, which generally require copies of past density matrices to be stored. The program was used to calculate 150 snapshots from each MD simulation in order to be able to estimate the electronic properties of the SAMs. Because periodic calculations are not yet possible, one simulation box without its Al_2O_3 slab was used for the snapshots, resulting in calculations for approximately 6,000 atoms.

ELECTRONIC PROPERTIES

One practical difficulty with MO calculations on many thousands of atoms is that the traditional analysis techniques (population analyses, individual MOs etc.) rapidly become too cumbersome for very large numbers of atoms. A more easily understood (and less space-demanding) technique is to calculate the values of informative local properties. The best known local properties are the electron density, which is usually used to construct an isodensity molecular surface [18] and the molecular electrostatic potential (MEP), [19] which indicates possible hot spots for Coulomb interactions. Local equivalents of the donor and acceptor levels are the local average ionization energy [20] and the local electron affinity, EA_L. [21, 22] These quantities are in essence density-weighted Koopmans theorem [23] ionization potential and electron affinity, respectively. An extensive review of local properties was given in an earlier article in this series [24].

For an n-type semiconductor, the most informative of these local properties is EA_L because it corresponds approximately to the acceptor level of the semiconductor and gives an easily understandable picture of electron transport in molecules and aggregates [25]. This technique allows us to visualize the space-resolved acceptor characteristics of the SAMs, as shown in figure 4.

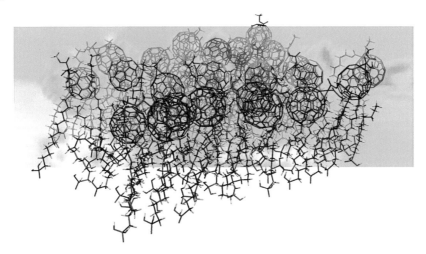

Figure 4. The local electron affinity (calculated using the AM1 Hamiltonian) for a diagonal vertical plane through a snapshot from the simulation illustrated in figure 2. The color coding runs from blue (most positive) to red (most negative). Electron affinities are defined as the ionization potential of the reduced species, so that the blue areas are those that accept electrons most readily.

Figure 4 shows a clear almost continuous blue band through the fullerene layer that indicates an almost uninterrupted path for electron transport through the upper layer of the SAM. However, this is only one snapshot. A complete analysis of the behavior of the device would require that the time-dependent behavior be analyzed using as many snapshots as possible from as long a simulation as possible. This is clearly not possible by visual inspection, so that more quantitative analysis techniques must be used.

ELECTRON-TRANSFER PATHS

Electron-transfer paths were determined by a simple Monte-Carlo procedure. "Electrons" (in reality simply points that are assigned the local electron affinity at their position as energy function) are subjected to a Metropolis Monte-Carlo procedure in which they are only allowed to move from the source towards the drain, not in the reverse direction. Performing many such simulations at different temperatures leads to many paths that do not cross to the drain (shown in blue in figure 5) and some "conducting paths (shown in gold) that reach the drain. Figure 5 shows the results of such simulations at kT = 215 mV.

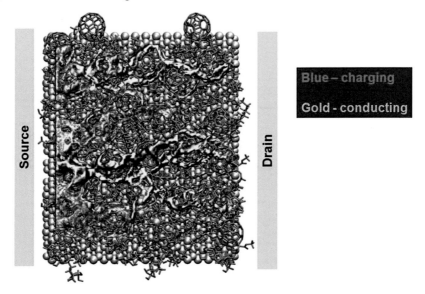

Figure 5. Electron-transfer paths through a snapshot of the mixed SAM. Blue paths are those that terminate in the SAM (and would therefore lead to accumulation of negative charge) and those in gold reach the drain electrode. The "source" and "drain" are in reality a little inside the edges of the SAM, as can be seen from the origins and ends of the paths. The Monte-Carlo simulation was carried out at kT = 215 mV.

The effect of the Monte-Carlo temperature on the paths can be seen in figure 6.

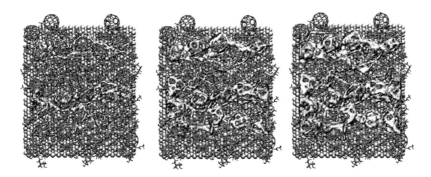

Figure 6. Electron-transfer paths through a snapshot of the mixed SAM at different "electronic" temperatures (kT = 215, 285 and 308 mV from left to right, respectively). The "source" and "drain" are in reality a little inside the edges of the SAM, as can be seen from the origins and ends of the paths.

The three sets of paths were found for the same MD snapshot at an "electronic" (= Monte Carlo) temperature of kT = 215, 285 and 308 mV from left to right, respectively. The paths shown in the left-hand (kT = 215 mV) plot correspond to the conducting paths shown in Figure 5. As expected, increasing the "electronic" temperature applied voltage) increases the number and cross sections of the "conducting" paths.

The paths shown in figure 5 are informative but at the moment do not take two factors into account. These are once more the conformational sampling, which would mean that hundreds or thousands of such simulations on different MD snapshots must be performed, and the fact that the "electrons" are uncharged and do not interact with each other.

"CONDUCTANCE" SIMULATIONS

A next step towards more realistic simulations of conductance in the SAMs is to include electron-electron interactions as point-charge Coulomb energies to be added to the local electron affinity. This is a very coarse approximation, but at least introduces the effect of electrons bound in the semiconductor layer. Figure 7 shows the results of one such simulation in which electrons are fed into the SAM from the source at a constant rate and removed from the simulation as soon as they reach the drain. The figure shows the situation after the simulation has come to equilibrium.

Clark, T. *et al.*

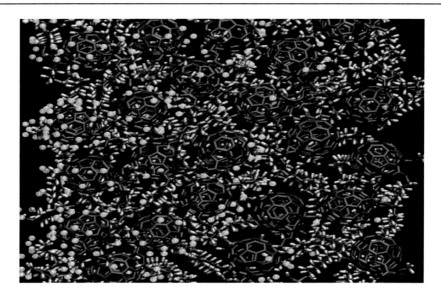

Figure 7. Snapshot of a Monte-Carlo simulation of conductance of classical point-charge electrons through the SAM. The electrons are shown as yellow balls and are free to move in all directions. The placing of the "electrodes" is the SAM as in figures 5 and 6.

Although the physical model used in the simulation is clearly not yet realistic, several features are in accord with our expectations. A potential across the SAM layer is built up by trapping electrons close to the source until the driving force provided by the electrostatic repulsion is sufficient to force the front-line electrons across the semiconductor layer to the drain. Once this potential has been built up, the simulation reaches a steady state in which electrons injected at the source result in others being removed at the drain. This model is also able to introduce a substantial negative charge into the semiconductor layer. Experiments suggest that this does indeed happen in fullerene-based SAMFETs and that the charge accumulation leads to significant hysteresis in the I/V curve. [12, 26] Even this very simple simulation suggests that the inter-fullerene van-der-Waals contacts serve as electron traps and that electrons are also trapped on the surfaces of the fullerene moieties. The latter is expected, but the high stability of electrons trapped between fullerenes was unexpected and is now being investigated using more sophisticated calculations.

SUMMARY AND OUTLOOK

The simulations described above are only the beginning. The model used for electron transport is primitive in the extreme and quantitative predictions of device characteristics such as I/V curves are not yet possible. Several significant effects must still be considered:

- The Monte-Carlo simulations have so far been performed on a single, static SAM-structure. Realistic simulations must couple the two time scales of atomic and electronic movement.

- The electrons have so far been treated as classical point charges. In reality, their traps are delocalized over a finite volume.

- Movements over barriers have also been treated classically. In reality, tunneling between the traps must be treated adequately.

- Perhaps the most difficult problem at the moment is to treat the interactions between the fullerenes in the presence of excess electrons correctly in the MD simulations. Reduced fullerenes do not always repel each other as they do in our classical model and may even form a covalent fullerene radical anion dimer [27]. These interactions must be taken into account, which is most easily achieved using direct QM/MM simulations, which will be computationally very expensive.

Nevertheless, much progress has been made in a research area for which conventional simulation techniques are often no longer adequate. Above all, the questions posed by the research goals must be answered by pushing existing techniques to their limits and making optimal use of computing facilities. This type of interdisciplinary research-driven method development over different time and length scales represents an important new direction in modern modeling and simulation.

ACKNOWLEDGMENTS

We thank Prof. Dirk Zahn for many discussions. This work was carried out in and financially supported by the Excellence Cluster *Engineering of Advanced Materials* funded by the *Deutsche Forschungsgemeinschaft*.

REFERENCES

[1]   See, for example, Raval, A., Eastwood, M.P., Dror, R.O., Shaw, D.E. (2012) Refinement of protein structure homology models via long, all-atom molecular dynamics simulations. *Proteins-Structure Function and Bioinformatics* **80**:2071 – 2079. doi: 10.1002/prot.24098.

[2]   Beierlein, F.R., Othersen, O.G., Lanig, H., Schneider, S., Clark T. (2006) Simulating FRET from Tryptophan: Is the Rotamer Model Correct? *J. Am. Chem. Soc.* **128**:5142 – 5152. doi: 10.1021/ja0584141.

[3]   Senn, H.M., Thiel W. (2009) QM/MM Methods for Biomolecular Systems. *Angew. Chemie Int. Ed. Engl.* **48**:1198 – 1229. doi: 10.1002/anie.200802019.

[4]   See, for example, Ochsenfeld C. (2009) Quantum Chemistry for Large Molecules: Linear-Scaling Mean-Field and Correlated Approaches. *AIP Conf. Proc.* **1108**:151 – 157. doi: 10.1063/1.3117123.

[5]   See, for example, Scuseria G.E. (1999) Linear Scaling Density Functional Calculations with Gaussian Orbitals. *J. Phys. Chem. A* **103**:4782 – 4790. doi: 10.1021/jp990629s.

[6]   Clark, T., Stewart J.J.P. (2011) NDDO-like Semiempirical Molecular Orbital Theory and its Application to Large Systems. In: Reimers, J.J. (ed.) Computational Methods for Large Systems. Wiley, Chichester, Chapter 8 (ISBN: 978 – 0-470 – 48788 – 4). doi: 10.1002/9780470930779.ch8.

[7]   Elstner, M., Hobza, P., Frauenheim, T., Suhai, S., Kaxiras, E. (2001) *J. Chem. Phys.* **114**:5149. doi: 10.1063/1.1329889.

[8]   Gräfenstein, J., Cremer, D. (2009) The self-interaction error and the description of non-dynamic electron correlation in density functional theory. *Theor. Chem. Acc.* **123**:171 – 182. doi: 10.1007/s00214-009-0545-9.

[9]   Yang, W. (1991) *Phys. Rev. Lett.* **66**:1438. doi: 10.1103/PhysRevLett.66.1438.

[10] Dixon, S.L., Merz, K.M. Jr. (1997) *J. Chem. Phys.* **107**:879; Ababoua, A., van der Vaart, A., Gogonea, V., Merz, K.M. Jr. (2007) *Biophys. Chem.* **125**:221.

[11] Stewart, J.J.P. (1996) *Int. J. Quant. Chem.* **58**:133. doi: 10.1002/(SICI)1097-461X(1996)58:2<133::AID-QUA2>3.0.CO;2-Z.

[12] Novak, M., Ebel, A., Meyer-Friedrichsen, T., Jedaa, A., Vieweg, B.F., Yang, G., Voitchovsky, K., Stellacci, F., Spiecker, E., Hirsch, A., Halik, M. (2010) Low-Voltage p- and n-Type Organic Self-Assembled Monolayer Field Effect Transistors. *Nano Letters* **11**:156 – 159.
doi: 10.1021/nl103200r.

[13] Smits, E.C.P., Mathijssen, S.G.J., van Hal, P.A., Setayesh, S., Geuns, T.C.T., Mutsaers, K.A.H.A., Cantatore, E., Wondergem, H.J., Werzer, O., Resel, R., Kemerink, M., Kirchmeyer, S., Muzafarov, A.M., Ponomarenko, S.A., de Boer, B., Blom, P.W.M., de Leeuw, D.M. (2008) Bottom-up organic integrated circuits. *Nature* **455**:956 – 959.
doi: 10.1038/nature07320.

[14] Wang, J., Wolf, R.M., Caldwell, J.W., Kollman, P.A., Case, D.A. (2004) Development and testing of a general amber force field. *J. Comput. Chem.* **25**:1157 – 1174.
doi: 10.1002/jcc.20035.

[15] Dewar, M.J.S., Thiel, W. (1977) *J. Am. Chem. Soc.* **99**:4899; Thiel, W. (1988) Encyclopedia of Computational Chemistry. In: Schleyer, P.v.R., Allinger, N.L., Clark, T., Gasteiger, J., Kollman, P.A., Schaefer III, H.F., Schreiner, P.R. (eds), Wiley, Chichester, 3, 1599.

[16] Thiel, W., Voityuk, A.A. (1992) *Theoret. Chim. Acta* **81**:391; Thiel, W., Voityuk, A.A. (1996) *Theoret. Chim. Acta* **93**:315; Thiel, W., Voityuk, A.A. (1994) *Int. J. Quant. Chem.* **44**:807; Thiel, W., Voityuk, A.A. (1994) *J. Mol. Struct.* **313**:141; Thiel, W., Voityuk, A.A. (1996) *J. Phys. Chem.* **100**:616.

[17] Hennemann, M., Clark, T. (2012) EMPIRE: A highly parallel semiempirical molecular orbital program. *J. Mol. Model.* (submitted).

[18] Bader, R.F.W. (1990) Atoms in Molecules: A Quantum Theory. Oxford University Press, Oxford.

[19] Politzer, P., Murray, J.S. (1998) Molecular electrostatic potentials and chemical reactivity. In: Lipkowitz, K., Boyd, R.B. (eds) *Rev. Comput. Chem., volume 2*, 273. VCH, New York.

[20] Sjoberg, P., Murray, J.S., Brinck, T., Politzer, P.A. (1990) *Can. J. Chem.* **68**:1440.
doi: 10.1139/v90-220.

[21] Ehresmann, B., Martin, B., Horn, A.H.C., Clark, T. (2003) Local molecular properties and their use in predicting reactivity. *J. Mol. Model.* **9**:342 – 347.
doi: 10.1007/s00894-003-0153-x.

[22]   Clark, T. (2010) The Local Electron Affinity for Non-Minimal Basis Sets. *J. Mol. Model.* **16**:1231 – 1238.
doi: 10.1007/s00894-009-0607-x.

[23]   Koopmans, T. (1933) Über die Zuordnung von Wellenfunktionen und Eigenwerten zu den Einzelnen Elektronen Eines Atoms. *Physica (Amsterdam)* **1**:104.
doi: 10.1016/S0031-8914(34)90011-2.

[24]   Clark, T., Byler, K.G., de Groot, M.J. (2008) Biological Communication via Molecular Surfaces. In: Molecular Interactions – Bringing Chemistry to Life. Proceedings of the International Beilstein Workshop (Eds. M.G. Hicks & C. Kettner), Bozen, Italy, May 15 – 19, 2006, Logos Verlag, Berlin, 129 – 146.

[25]   Atienza, C., Martin, N., Wielepolski, M., Haworth, N., Clark, T., Guldi, D. (2006) Tuning electron transfer through *p*-phenyleneethylene molecular wires. *J. Chem. Soc. Chem. Commun* 3202 – 3204.
doi: 10.1039/b603149h.

[26]   Burkhardt, M., Jedaa, A., Novak, M., Ebel, A., Voïtchovsky, K., Stellacci, F., Hirsch, A., Halik, M. (2010) Concept of a Molecular Charge Storage Dielectric Layer for Organic Thin-Film Memory Transistors. *Advanced Materials* **22**:2525 – 2528.
doi: 10.1002/adma.201000030.

[27]   Komatsu, K., Wang, G.-W., Murata, Y., Tanaka, T., Fujiwara, K., Yamamoto, K., Saunders, M. (1998) Mechanochemical Synthesis and Characterization of the Fullerene Dimer C 120. *J. Org. Chem.* **63**:9358 – 9366.
doi: 10.1021/jo981319t.

Beilstein-Institut

Beilstein Bozen Symposium on Molecular Engineering and Control
May 14th – 18th, 2012, Prien (Chiemsee), Germany

151

NON EQUILIBRIUM STRUCTURED POLYNUCLEAR-METAL-OXIDE ASSEMBLIES

LEROY CRONIN

School of Chemistry, University of Glasgow, WestCHEM,
Glasgow, G12 8QQ, United Kingdom

E-MAIL: Lee.Cronin@glasgow.ac.uk

Received: 5th February 2013 / Published: 13th December 2013

ABSTRACT

One pot reactions are deceptively simple systems often yielding complex mixtures of compounds, nanomolecular self-assembled architectures and intricate reaction networks of interconnected mutually dependent processes. As such, the elucidation of mechanism and various reaction pathways can be hard if not impossible to deduce. Herein, I show how by moving a 'one-pot' reaction from the time domain into a flow-system, the time domain translates into distance and flow rate thereby allowing monitoring and control of one-pot reactions in new ways; for example by changing the tube length/diameter. Three types of flow system are presented: (i) a system for the trapping of an intermediate host guest complex responsible for the formation of the giant wheel cluster which is the major component of molybdenum blue; (ii) a linear flow system array for the scale up of inorganic clusters; (iii) a networked reactor system which allowed the combination of multiple one-pot conditions in a single system allowing the discovery of a fundamental new class of inorganic cluster not accessible by any other means. I also briefly describe our recent work on the growth of inorganic tubules and our 3D printed 'reactionware' for the fabrication of bespoke flow-systems at a fraction of the cost of commercial systems and also show how the ability to configure the systems in new ways leads to new science.

INTRODUCTION

The discovery and synthesis of complex chemical nanostructures, both molecular and supra-molecular, is both time consuming and limited by lack of reproducibility, amount of material, and dependence on initial starting conditions. However, flow system approaches, utilized a lot in organic synthesis, have not been utilized for the synthesis of supramolecular and nanomolecular chemistry. Continuous flow processing in synthetic organic chemistry has many benefits and has been well researched and documented in recent years [1, 2]. Key advantages include higher heat transfer efficiencies, rapid and homogeneous mixing and this can lead to higher reaction rates with less side products, and consequently higher yields and selectivities [3, 4]. Continuous flow process techniques have also proved useful in a limited number of hard nanomaterials but examples are limited mainly to the production of metallic and semiconductor nanoparticles and quantum dots [5]. Other major areas of interest in inorganic chemistry, such as coordination compounds and clusters such as polyoxometalates (POMs) [6, 7] (especially those with interesting optical, redox, and magnetic properties etc) [8 – 12], typically utilize batch syntheses and purification *via* crystallization. This means that the main role of the synthetic chemist is to systematically screen the reaction conditions for the reproduction and scale up of novel architectures must therefore cover a large area of synthetic space as it not only has to achieve conditions suitable for the target formation, but target crystallization as well [13]. Large number reaction arrays are therefore an inherent aspect in this process, which can be an extremely laborious and time consuming task when working solely under batch conditions, especially when exploring delicate multi-parameter self-assembly reactions aiming to produce supramolecular architectures. Indeed, the size of the parameter space can be so vast that even after chance 'batch-discovery' the re-discovery and scale up of the process to produce more than a few crystals of the product can be almost impossible on a limited time-scale. As such, this presents a critical bottleneck to the reliable synthesis of inorganic macromolecules with scientifically and technologically important physical properties, where the lack of phase-pure material prevents a rigorous investigation of the physical properties or exploitation of the properties in real world devices and applications.

TEMPLATING GIGANTIC MOLECULAR ASSEMBLY

In one of our first flow system devices we aimed to probe the assembly mechanism of gigantic molecular nanoparticles such as the wheel-shaped $\{Mo_{154-x}\}$ ($x = 0 - 14$) (the mixed-valence molybdenum blue (MB) type) cluster family [14] which have interesting physical and chemical properties, which arises from their molecular nature, nanoscale size, and electronic properties [15]. Our idea was that the inorganic MB-macrocycles can only be formed around a templating cluster, which is built of the same building blocks as the macrocycle itself and therefore acts as a building blocks feedstock, during the process of macrocycle assembly but this was hard to prove as the MB type clusters are synthesised by the reduction of an acidic molybdate solution (pH $1 - 3$) under static one-pot conditions

[16–20]. Under these conditions, the self-assembly of reactive $[Mo_xO_y]$-type building blocks (in the final structures mostly given as $\{Mo_1\}$, $\{Mo_2\}$, $\{(Mo)Mo_5\}$ and $\{Mo_8\}$, see Fig. 1) gives rise to $\{Mo_{154-x}\}$ (x = 0 – 14), which can be rationalized formally as assemblies of the above mentioned building blocks. Yet the reaction conditions effectively concealed the template required for the self-assembly process. We were able to deduce this by using a controllable dynamic synthetic procedure in a flow system such that it is possible to adjust the three input variables (pH, concentration of molybdate and reducing agent) which control the synthesis of the molecular nanosized-wheels in real time compared to the static synthetic approach, see Fig. 1.

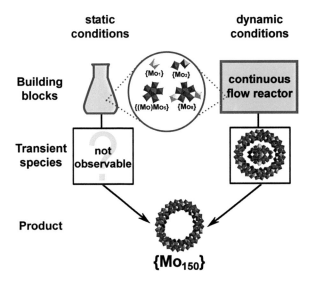

Figure 1. Polyhedral representation of the basic building units (present in the final structure) which govern the (soluble) molybdenum-oxide based, i.e. MB chemistry with giant clusters as well as schematic representation of the static and dynamic synthetic approach to assemble the 2.6 nm diameter molecular nano wheel. The transient species, the molecular wheel with the template 'X' is observed in a flow system crystallizer.

By utilising a flow-system, as opposed to the classical one-pot reaction, we are able to maintain the concentration of the transient by controlling the degree of reduction of the polyoxomolybdate reaction solution (see Fig. 2). As such the assembly of the MB architecture under controlled reduction conditions leads to the trapping of a $\{Mo_{150}\}$ wheel with a $\{Mo_{36}\}$ cluster template that is bound to the central cavity of the ring species via sodium cations. This host guest complex shows features indicative of an intermediate electronic and structural state, and we conclude that the $\{Mo_{36}\}$ cluster acts as the key template in the formation of the MB ring. The host guest complex is isolated as the crystalline compound $Na_{22}[Mo^{VI}_{36}O_{112}(H_2O)_{16}] \subset [Mo^{VI}_{130}Mo^{V}_{20}O_{442}(OH)_{10}(H_2O)_{60}] \cdot 180\,H_2O$ in the flow-system by the reduction of an aqueous acidic solution of $Na_2MoO_4 \cdot 2\,H_2O$ with $Na_2S_2O_4$ under continuous addition of HNO_3 in a gram yield. The use of nitric acid in the flow system is important as it has the dual role of a proton source, and as an oxidant causing incomplete reduction of the wheel. The "normal", i.e. complete/symmetrical wheel $\{Mo_{154}\}$ has 14 two-electron reduced compartments (i.e. with a total of 28 4d electrons) but here only 10 of the 14 compartments are two-electron reduced with the wheel being a total of 20-electron reduced.

Figure 2. Photograph of the flow system with the reduced molybdenum blue and a polyhedral representation of the templated $\{Mo_{36}\} \subset \{Mo_{150}\}$ ring which crystallizes from the flow reactor.

'ONE-POT' REACTIONS OF POLYOXOMETALATE SYNTHESIS IN LINEAR FLOW SYSTEM ARRAYS

In this approach we took the multiple input 'ingredients' normally added to the 'one-pot' reaction systems and made up each ingredient as a separate flow input into a reaction chamber where the number of inputs into the reaction chamber is normally equal to the number of distinct reagents used in the stand alone 'one-pot' reaction. By using a continuous flow processing with batch crystallization over a set array, we were able to devise a new efficient method for creating large number reaction arrays for rapidly scanning large areas of reaction parameter space to increase the probability of discovering new materials. Also, this flow based approach for the generation of multiple batch reactions can be used for the continuous production of identical batch reactions required for scale-up of the isolated materials. We were able to validate this approach by setting up a multiple pump reactor system for the synthesis of a range polyoxomolybdates of various sizes and structural complexity, $\{Mo_x\}$ compounds shown in figure 3 [13–16]. The setup utilized twelve programmable syringe pumps, although this is extendable to fifteen in our system, and a PC interface was used for controlling the pumps (Fig. 1.).

The reagent set chosen for POM synthesis consisted of distilled deionized water for dilution, 2.5 M $Na_2MoO_4 \cdot 2 H_2O$ as the molybdenum source, three acid sources (5.0 M HCl, 1.0 M H_2SO_4, and 50% AcOH), 4.0 M $AcO(NH_4)$, and two sources of reducing agent, 0.25 M $Na_2S_2O_4$ and saturated (0.23 M) $N_2H_2 \cdot H_2SO_4$. For the simplest POM target, compound $\{Mo_{36}\}$, only three of the twelve pumps were required to incrementally vary the relative flow rates of the water, molybdate and HCl stock solutions; for the $\{Mo_{154}\}$, $\{Mo_{132}\}$, $\{Mo_{102}\}$, and $\{Mo_{368}\}$ compounds up to five pumps were required to supply the additional reducing agent and buffer stocks [21]. In the example of the synthesis of the $\{Mo_{36}\}$ compound, this involves acidifying an aqueous solution of sodium molybdate, which subsequently precipitated crystals of the target compound, and we prepared a pump setup that could repeat this screening process. With $\{Mo_{36}\}$ for our "screening array", the pumps were programmed to run at a range of flow rates, incrementally increasing both the relative ratio of acid to molybdate and the overall reagent concentrations (two key parameters of POM formation and crystallization) throughout the experimental run. In this case variation of just these two parameters resulted in the creation of fifty distinct reaction batches with the potential to crystallise the $\{Mo_{36}\}$ target and crucially the successful systems could be repeated many times reliably to allow the isolation of many grams of material if required.

'ONE-POT' REACTIONS OF POLYOXOMETALATE SYNTHESIS IN NETWORKED REACTION SYSTEMS

One important problem is that 'one-pot' reactions mask a vast and complex range of intricate self-assembly processes that must invariably occur in solution, and it is therefore difficult to predict or control the assembly process. This issue is not limited to polyoxometalates, but extends to a vast number of other chemical systems e.g. supramolecular, nanoparticles, DCLs and coordination chemistry. To go beyond our previous work we theorised that it could be possible to systematically explore the systems by doing combinatorial reactions one-by-one, but this only probes the combinatorial space in a very limited sense. However, the rapid screening and integration of a large number of 'one-pot' reactions would be transformative since this would allow both the compositional and time dependent space to be integrated, and 'networked', and we recently proposed this and published the first example of such a set up experimentally, see figure 4 [22].

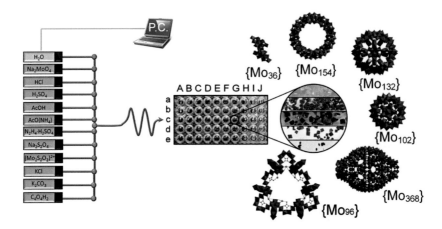

Figure 3. Reaction input parameters were altered to generate reaction arrays that were screened for the crystallization of polyoxomolybdate targets. A schematic of the pump setup and reagent inputs (left) leads to a representative image of the 5×10 reaction array outputs (centre). During the screening process for each target, crystal batches were obtained in a handful of reactions where the flow conditions produced the reaction conditions required for successful crystallization. Images of crystal batches and polyhedral structural representations for the target polyoxomolybdates are shown (right). The truncated molecular formulae in parentheses represent the following complete formulae: $\{Mo_{36}\} = Na_8[Mo_{36}O_{112}(H_2O)_{16}] \cdot 58\,H_2O$; $\{Mo_{154}\} = Na_{15}[Mo^{VI}_{126}Mo^{V}_{28}O_{462}H_{14}(H_2O)_{70}]_{0.5}[Mo^{VI}_{124}Mo^{V}_{28}O_{457}H_{14}(H_2O)_{68}]_{0.5} \cdot ca.\ 400\,H_2O$; $\{Mo_{132}\} = (NH_4)_{42}[Mo^{VI}_{72}Mo^{V}_{60}O_{372}(CH_3COO)_{30}\ (H_2O)_{72}] \cdot ca.\ 300\,H_2O \cdot ca.\ 10CH_3COONH_4$; $\{Mo_{102}\} = Na_{12}[Mo^{VI}_{72}Mo^{V}_{30}O_{282}(SO_4)_{12}(H_2O)_{78}] \cdot ca.\ 280\,H_2O$; $\{Mo_{368}\} = Na_{48}[H_xMo_{368}O_{1032}(H_2O)_{240}(SO_4)_{48}] \cdot ca.\ 1000\,H_2O$; and $\{Mo_{96}\} = (N(CH_3)_4)_6K_{30}\{\ [(Mo_2O_2S_2)_3(OH)_4(C_4O_4)]_9[(Mo_2O_2S_2)_2(OH)_2(C_4O_4)]_3 (Mo_5O_{18})_6\} \cdot 132\,H_2O$.

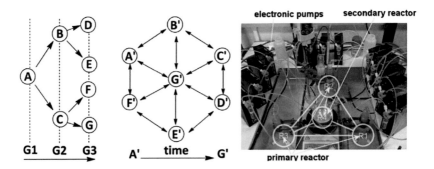

Figure 4. Left: comparison between conventional parameter space, a, (i. e. generations G1, G2 and G3) and networked multiple parameter, b, screenings, in X or X' 'one-pot' reactions (where X = A to G). Right: photograph of the physical networked 'one-pot' reactor array, c.

Thus the development of a networked 'one-pot' reaction array (Fig. 4c) should be of fundamental importance since linking multiple complex assembly processes, such as those found in one-pot systems, provides potential not only for the reproducible assembly of complex nanostructures, but also allows the systematic combination of one-pot reactions of similar systems as a function of time or composition permitting the exploration of virtual libraries of building blocks. Potentially this could lead to the control of assembly at the molecular level using 'macro-control' in a series of 'one-pot' reactions connected in a flow system [22].

By designing and setting up a Networked Reactor System (NRS) for the discovery of polyoxometalate clusters, we applied this approach to the synthesis of an unknown family of metal-containing isopolyoxotungstates (iso-POTs) in presence of templating transition metals such as Co^{2+}, see figure 5, by screening networks of 'one-pot' reactions. This shows that the NRS approach can lead to the discovery of new clusters in a reproducible way allowing one-pot reactions to be probed or expanded over a number of reaction vessels, rather than relying on one single vessel. As such, the use of the NRS leads to the discovery of a chain-linked iso-POT $\{(DMAH)_6[H_4CoW_{11}O_{39}]\bullet 6\,H_2O\}_n \equiv \{W_{11}Co\}_n$, a Co-trapped iso-POT $Na_4(DMAH)_{10}[H_4CoW_{22}O_{76}(H_2O)_2]\bullet 20\,H_2O \equiv \{W_{22}Co\}$ and, $Na_{16}(DMAH)_{72}$ $[H_{16}Co_8W_{200}O_{660}(H_2O)_{40}]\bullet ca600\,H_2O \equiv \{W_{200}Co_8\}$ which is over 4 nm in diameter and it represents the largest discrete polyoxotungstate cluster so far characterized [22]. This cluster is formed uniquely in the NRS since several different one-pot reaction processes can be set up independently and mixed together leading to the interconnection of building blocks synthesized in the network of reactors which are then linked to yield the final cluster compounds.

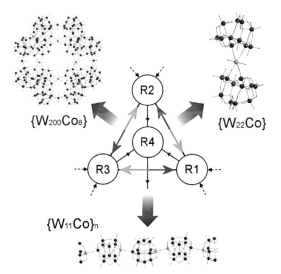

Figure 5. Scheme of Networked Reactor System (red and green arrows show the clockwise and anticlockwise circulation of the 'one-pot' reactions) and the new structures are highlighted around the triangle-shaped networked reactor system: Crystal structures of compounds $\{W_{11}Co\}_n$, $\{W_{22}Co\}$ and $\{W_{200}Co_8\}$ are shown in ball and stick mode. Colour scheme: W purple, Co cyan, O red.

The discovery of the $\{W_{200}\}$ is accomplished uniquely in the NRS which is interesting since this opens the way for nanoscale control using macro-scale parameters. This is because the NRS is designed to combine two aspects: the synthesis of new compounds by linking separate 'one-pot' reactions each containing unique building blocks, (BBs) followed by the mixing of these individual 'one-pot' reactions i.e. moving the regents from reactor to reactor. Thus the NRS allows control in both reaction in both *time* and *space* (by comparison we consider that normal 'one-pot' reactions only search in time). As such three primary reactors, each with two external reagent inputs, are connected together in a triangular arrangement with a central secondary reactor (connecting to all three primary reactors), defining the simplest implementation of the NRS. As the NRS has a high connectivity, this allows a wide range of multiple mixing pathways in which the reagents can move from one flask to another (i.e. anti-clockwise $R1 \rightarrow R2 \rightarrow R3 \rightarrow R4$ or clockwise $R1 \rightarrow R3 \rightarrow R2 \rightarrow R4$). This allows the recycling and re-feeding processes according to $(R1 \rightarrow R2 \rightarrow R3)_n$ (n = number of cycles) depending of standard flow parameters in the NRS. In contrast to a linear setup, the NRS allows many different reagent inputs to be accommodated in separate reactors. Moreover, the system can allow both the screening and automation of the syntheses over a range of different clusters by selecting the reaction and flow parameters (flow rates, pH, initial volumes, etc.) in a highly automated, controlled and reproducible manner.

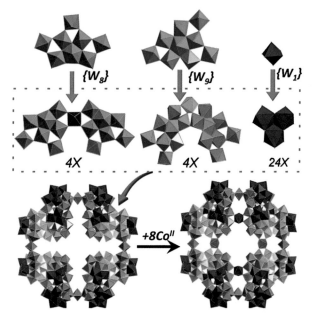

Figure 6. Representation of the crystal structure of the $\{W_{200}\}$. The three principal building blocks are represented at the top, where $\{W_8\}$ and $\{W_9\}$ are derived from $\{W(W)_5\}$ pentagonal unit. Following to principal BBs, the secondary BBs are represented in the middle section, as result of condensation of primary building blocks. Finally, the compound is completed by the complexation of 8 cobalt ions.

The gigantic isopolyanion compound $\{W_{200}Co_8\}$, is a saddle-shaped structure and contains unusual pentagonal units and crystallizes as a hydrated sodium and dimethylammonium salt of $\{H_{16}Co_8W_{200}O_{660}(H_2O)_{40}\}^{88-}$ and single crystal X-ray diffraction shows the crystals belonging tetragonal system with space group of $P4_2/nmc$. The cluster itself has an approximate D_{2d} symmetry and the building blocks are highlighted in figure 6.

Thus our idea of the networked reactor system (NRS) has been realised where multiple one-pot connected reactions are screened, the reaction variables explored, and automation of the syntheses of three compounds was achieved. The potential of the NRS methodology is transformative due to the ability to explore one-pot reactions as configurable modules, and to explore different mixing and reaction conditions in a programmed and sequential way (stepwise process) as well as allowing the combination of building block libraries that could not coexist in classical one-pot reactions. This is because the NRS allowed the combination of pH adjustment/UV monitoring in real time, thus confirming different local experimental conditions in each reactor within the system. This feature makes the NRS potentially very useful to explore other combinations of initial reagents, to study reaction mechanisms and self-assembly reactions in other areas of chemistry (i.e. coordination chemistry or design of metal-organic frameworks). As such, we demonstrate 'macroscale' control of the assembly of polyoxometalates for the first time, and this builds on our observations of 'microscale' control of assembly and opens perspectives to utilise the approach here in exploring 'assembly-isomers' of polyoxometalates in the NRS, as well as providing radically new structures (very high charge).

GROWTH OF INORGANIC MICROTUBES FROM POLYOXOMETALATES

Initially reported by us in 2009, the growth of micron-scale hollow tubes from polyoxometalate (POM) materials undergoing cation exchange with bulky cations in aqueous solution has now been shown as a general phenomenon for POMs within a critical solubility range. While it is not a classical chemical garden process, there are many similarities in the growth mechanism [23 – 27]. Although POMs have much in common with bulk transition metal oxides, their molecular nature means they have a vast structural diversity with many applications as redox, catalytically active and responsive nanoscale materials. Tube growth has been demonstrated with a wide range of different cations including several dihydro-imidazo-phenanthridinium (DIP) compounds, 3,7-bis(dimethylamino)-phenothiazin-5-ium chloride (methylene blue), polymeric poly(N-[3-(dimethylamino)propyl]methacrylamide) and even the complex $Ru^{II}(bipy)_3(BF_4)_2$ (bipy = 2,2'-bipyridyl). When dissolved anionic POM fragments, with their associated small cations, come into contact with bulky cations in solution, ion exchange occurs and the resulting aggregates grow sufficiently large to become insoluble. Due to the charge to size ratios of the molecules involved, it is often not favourable to make a fully charge balanced species, and so the material tends to be 'sticky', which results in continued aggregation to form precipitation membranes. When the dissolved POM material is introduced to the cation solution via a small aperture this tendency results in the formation of a continuous hollow (tube) structure, see figure 7.

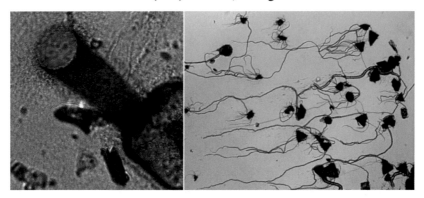

Figure 7. Growth of a microtube (approx. 50 μm diameter) growing on a glass surface. $(C_4H_{10}NO)_{40}[W_{72}Mn_{12}O_{268}Si_7]\cdot48\,H_2O$ POM in N-methyl dihydroimidazolphenanthridinium bromide solution. Right: Growth of several microtubes (approx. 25 – 75 μm diameter) under bulk flow conditions showing the alignment of the growth direction to the flow.

The tube continues to grow from the open end until the source of POM material is exhausted, a shorter exit route is provided (e. g. from a rupture in the tube closer to the source), or the cation concentration becomes too low for aggregation. In general, the POM solutions are denser than the cation solutions, and so the microtubes grow along the bottom surface of their sample container. However, when the solution densities are closely matched or the sample volume is very small, the tubes will 'climb' along the sample container walls or any other introduced obstacles, or can be seen to grow vertically into the solution.

In order to make useful patterns or devices from POM microtubes, their characteristics must be controllable. As the POM material flows from the end of the growing tube, the rate of aggregation is strongly influenced by the local availability (concentration) of cations and the rate of outflow. When the rate or concentration is increased, the material begins to aggregate more rapidly and the resulting tube narrows. The opposite is true if the rate or concentration is reduced. In the case of tubes growing from a crystal, the outflow is essentially fixed, as the surface area of the membrane (and thus the rate at which water can cross it) does not change. Since the POM material is of low solubility, the POM solution within the membrane and tube will be at saturation. This is not the case when the tube is grown by microinjection, since the pressure can be varied and this allows a direct control over the diameter.

As the POM material is ejected from the growing tube, it is influenced by any liquid flow in the sample such that tubes will always grow along the direction of flow. In a bulk sample, the use of electrodes to set up convection currents allows the direction of the growing tubes to be controlled. However, this does not allow the production of any useful network or tube-based device as the control is exerted on all the tubes simultaneously (see figure 8 RHS). By using a dye molecule in the cation solution, a focused laser spot can be used to create a localized flow; at the laser spot, the solution is heated and this creates an upward flow. To relieve the low pressure that this generates, there is a localized flow towards the focused spot. Any growing microtubes in the vicinity are then pulled towards the spot and by moving the spot the growth direction can be controlled. This means that an individual tube can be steered reliably and independently. By coupling the laser optics to a spatial light modulator (SLM) in a setup more commonly used as 'optical tweezers', the laser light can be split into multiple foci which can be used to control different growing tubes independently within the same sample. Growing a specific structure requires accurate positioning of the laser spots by a user who can react to the progress of the system, and this depends critically on the computer interface used to control the SLM. To achieve this, the microscope image is displayed on a multi-touch tablet (Apple iPad), along with markers representing the laser spots. These markers can be dragged around, moving the laser spots to follow the user's fingers, so that multi-touch gestures allow the growing microtubes to be controlled (examples of structures in figure 8 left). The laser heating also allows tube walls to be deliberately ruptured, producing branch points, or pre-loaded capillaries to be un-blocked in-situ so that extra tubes can be initiated on demand during 'device' construction. It is also possible to automate the movement of the laser spot with image analysis and feedback such that the computer can control the laser spot to steer a microtube into a predefined pattern as it grows (see Fig. 8 right).

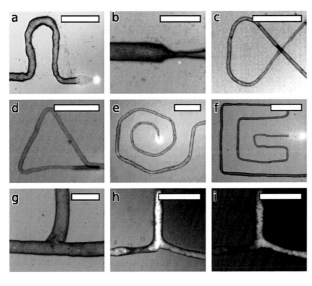

Figure 8. The holographic system can produce devices using a number of basic elements of the inorganic micro networks: **(a)** a sharp bend (scale bar 100 μm), **(b)** a change in diameter (scale bar 100 μm), **(c)** tubes crossing over one another (scale bar 500 μm), **(d)** a triangle motif (scale bar 250 μm), **(e)** a spiral pattern (scale bar 500 μm), **(f)** a nested pattern (scale bar 500 μm), **(g)** a 'T' junction produced by puncturing a growing tube (scale bar 100 μm), **(h)** a 'Y' junction produced by merging two tubes (scale bar 250 μm) and **(i)** a similar 'Y' junction where green fluorescent dye (Fluorescein) is introduced into one of the tubes (scale bars = 500 μm).

3D Printed 'Reactionware' as Configurable Flow System Devices

Traditionally, the fabrication of flow devices and their interfacing with in-line analysis is complicated and expensive as micro-scale fluidic devices have been required. We have been working on a convergent approach developing 3D-printed milli-scale flow devices, or tailored "reactionware" for chemical reactions and in-line analysis [28, 29]. The use of three-dimensional (3D) printing bypasses sophisticated manufacturing centres, and promises to revolutionise every part of the way that materials are turned into functional devices, from design through to operation, with 3D printing producing bespoke, low-cost appliances which previously required dedicated facilities. 3D printing is a cheap chemical discovery tool which combines the disciplines of synthetic chemistry and chemical engineering in a re-configurable and highly accessible format. The use of freely available CAD software and the rapid fabrication that comes with 3D printing, allows for the design and production of specific devices tailored to the intended chemical reaction. The high surface area-to-volume-ratio, precise control of volume and manipulation of reaction environment results in strict control of the final device, and the subsequent reactions carried out.

We have previously demonstrated the versatility and configurability of reusable and bespoke reactionware, where 3D printing was used to initiate chemical reactions by printing reagents directly into a 3D 'reactionware' matrix [29, 30]. We have also presented how 3D printing can be used to make intricate micro- and milli- scale reactionware for organic, inorganic and materials syntheses, offering significant freedom to design bespoke reactors in terms of residence time, mixing points, inlets and outlets, etc., see figure 9. For example in the reactor shown in figure 9, it is possible to synthesise Mo-blue as in the much larger and more difficult to control flow system shown in figure 2.

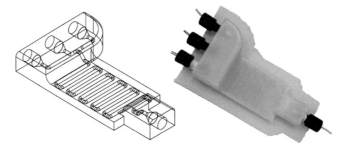

Figure 9. On the left is a schematic presentation of the .stl file, whilst on the right is the device with screw fittings and connected with 1/16th inch tubing. Methylene Blue and Rhodamine B are being pumped through the device, which allows for the inner tube-path to be rendered visible. A section consisting of only Methylene Blue can be seen at the front, followed by a stronger purple band, which is obtained from the successful mixing of Rhodamine B and Methylene Blue.

CONCLUSIONS

Molecular metal oxides based upon polyoxometalates are incredibly interesting compounds but they are extremely complex and the mechanism of assembly leading to real architectural control is far from reach under normal reaction conditions. To combat this problem we have developed new reaction approaches which allow the exploration of the self-assembly of complex molecular, supramolecular and nanomolecular species in fundamentally different ways to that traditionally envisaged by synthetic chemists. In particular the use of linear, branched, and networked flow systems to probe the fundamentals of molecular assembly is a new approach which promises to allow us to program the assembly of complex architectures by combining concepts from supramolecular chemistry, with reaction kinetics and non-equilibrium processing. Further, by developing optical guidance systems it is possible to grow physical structures of the metal oxides using computer control on the micron scale and finally the use of 3D printed 'reactionware' gives the ability to design new reactor systems with in-line analysis and new control approaches is now within reach.

ACKNOWLEDGEMENTS

I would like to acknowledge all my research group members, collaborators and funders especially the EPSRC and the University of Glasgow who have helped make all this research work possible over the past decade.

REFERENCES

[1]   *Chemical Reactions and Processes under Flow Conditions*; Luis, S.V., Garcia-Verdugo, E., Eds.; Royal Society Chemistry: Cambridge, (2010).

[2]   Seeberger, P.H. (2009) Organic synthesis: scavengers in full flow. *Nat. Chem.* **1**:258–260.
doi: 10.1038/nchem.267.

[3]   Wegner, J., Ceylan, S. & Kirschning, A. (2011) Ten key issues in modern flow chemistry. *Chem. Commun.* **47**:4583–4592.
doi: 10.1039/c0cc05060a.

[4]   Hartman, R.L., McMullen, J.P. & Jensen, K.F. (2011) Deciding whether to go with the flow: evaluating the merits of flow reactors for synthesis. *Angew. Chem. Int. Ed.* **50**:7502–7519.
doi: 10.1002/anie.201004637.

[5]   Abou-Hassan, A., Sandre, O. & Cabuil, V. (2010) Microfluidics in inorganic chemistry. *Angew. Chem. Int. Ed.* **49**:6268–6286.
doi: 10.1002/anie.200904285.

[6]   Long, D.-L., Burkholder, E. & Cronin, L. (2007) Polyoxometalate clusters, nano-structures and materials: from self assembly to designer materials and devices. *Chem. Soc. Rev.* **36**:105 – 121.
        doi: 10.1039/b502666k.

[7]   Long, D.-L., Tsunashima, R. & Cronin, L. (2010) Polyoxometalates: building blocks for functional nanoscale systems. *Angew. Chem. Int. Ed.* **49**:1736 – 1758.
        doi: 10.1002/anie.200902483.

[8]   Evangelisti, M. & Brechin, E.K. (2010) Recipes for enhanced molecular cooling. *Dalton Trans.* **39**:4672 – 4676.
        doi: 10.1039/b926030g.

[9]   Inglis, R., Milios, C.J., Jones, L.F., Piligkos, S. & Brechin, E.K. (2012) Twisted molecular magnets. *Chem. Commun.* **48**:181 – 190.
        doi: 10.1039/c1cc13558a.

[10]   Moushi, E.E. *et al.* (2010) Inducing single-molecule magnetism in a family of loop-of-loops aggregates: heterometallic $Mn_{40}Na_4$ clusters and the homometallic Mn_{44} analogue. *J. Am. Chem. Soc.* **132**:16146 – 16155.
        doi: 10.1021/ja106666h.

[11]   Murrie, M. (2010) Cobalt(II) single-molecule magnets. *Chem. Soc. Rev.* **39**:1986 – 1995.
        doi: 10.1039/b913279c.

[12]   Wang, X.-Y., Avendano, C. & Dunbar, K.R. (2010) Molecular magnetic materials based on 4 d and 5 d transition metals. *Chem. Soc. Rev.* **40**:3213 – 3238.
        doi: 10.1039/c0cs00188k.

[13]   Richmond, C.J., Miras, H.N., de la Oliva, A.R., Zang, H., Sans, V., Paramonov, L., Makatsoris, C., Inglis, R., Brechin, E.K., Long, D.-L., & Cronin, L. (2012) A flow-system array for the discovery and scale up of inorganic clusters. *Nature Chem.* **4**:1037 – 1043.
        doi: 10.1038/nchem.1489.

[14]   Müller, A. *et al.* (1999) Rapid and simple isolation of the crystalline molybdenum-blue compounds with discrete and linked nanosized ring-shaped anions: $Na_{15}[Mo^{VI}_{126}Mo^{V}_{28}O_{462}H_{14}(H_2O)_{70}]_{0.5}$ $[Mo^{VI}_{124}Mo^{V}_{28}O_{457}H_{14}(H_2O)_{68}]_{0.5}$ · ca. 400 H_2O and $Na_{22}[Mo^{VI}_{118}Mo^{V}_{28}O_{442}H_{14}(H_2O)_{58}]$ · ca. 250 H_2O. *Z. Anorg. Allg. Chem.* **625**:1187 – 1192.
        doi: 10.1002/(SICI)1521-3749(199907)625:7<1187::AID-ZAAC1187>3.0.CO;2-#.

[15]   Miras, H. N. *et al.* (2012) Unveiling the transient template in the self assembly of a molecular oxide nano-wheel. *Science* **327**:72 – 74.
        doi: 10.1126/science.1181735.

[16] Müller, A., Krickemeyer, E., Bögge, H., Schmidtmann, M. & Peters, F. (1998) Organizational forms of matter: an inorganic super fullerene and keplerate based on molybdenum oxide. *Angew. Chem. Int. Ed.* **37**:3359 – 3363. doi: 10.1002/(SICI)1521-3773(19981231)37:24<3359::AID-ANIE3359>3.0.CO;2-J.

[17] Krebs, B., Stiller, S., Tytko, K.H. & Mehmke, J. (1991) Structure and bonding in the high-molecular-weight isopolymolybdate ion, $(Mo_{36}O_{112}(H_2O)_{16})^{8-}$ – the crystal structure of $Na_8(MO_{36}O_{112}(H_2O)_{16}) \cdot 58\,H_2O$. *Eur. J. Solid State Inorg. Chem.* **28**:883 – 903.

[18] Müller, A., Beckmann, E., Bögge, H., Schmidtmann, M. & Dress, A. (2002) Inorganic chemistry goes protein size: a Mo_{368} nano-hedgehog initiating nanochemistry by symmetry breaking. *Angew. Chem. Int. Ed.* **41**:1162 – 1167. doi: 10.1002/1521-3773(20020402)41:7<1162::AID-ANIE1162>3.0.CO;2-8.

[19] Müller, A. *et al.* (1995) $[Mo_{154}(NO)_{14}O_{420}(OH)_{28}(H_2O)_{70}]^{(25\ \pm\ 5)-}$: A water-soluble big wheel with more than 700 atoms and a relative molecular mass of about 24000. *Angew. Chem. Int. Ed.* **34**:2122 – 2124. doi: 10.1002/anie.199521221.

[20] Miras, H.N., Richmond, C.J., Long, D.-L. & Cronin, L. (2012) Solution-phase monitoring of the structural evolution of a molybdenum blue nanoring. *J. Am. Chem. Soc.* **134**:3816 – 3824. doi: 10.1021/ja210206z.

[21] Miras, H.N., Yan, J., Long, D.-L. & Cronin, L. (2012) Engineering polyoxometalates with emergent properties. *Chem. Soc. Rev.* **41**:7403 – 7430. doi: 10.1039/c2cs35190k.

[22] de la Oliva, A.R., Sans, V., Miras, H.N., Yan, J., Zang, H., Richmond, C.J., Long, D.-L. & Cronin, L. (2012) Assembly of a Gigantic Polyoxometalate Cluster $\{W_{200}Co_8O_{660}\}$ in a Networked Reactor System. *Angew. Chem. Int. Ed.* **51**:12759 – 12762. doi: 10.1002/anie.201206572.

[23] Ritchie, C., Cooper, G.J.T., Song, Y.-F., Streb, C., Yin, H., Parenty, A.D.C., MacLaren, D.A., & Cronin, L. (2009) Spontaneous Assembly and Real-Time Growth of Micron-Scale Tubular Structures from Polyoxometalate-Based Inorganic Solids. *Nature Chem.* **1**:47 – 52. doi: 10.1038/nchem.113.

[24] Cooper, G.J.T. & Cronin, L. (2009) Real-Time Direction Control of Self Fabricating Polyoxometalate-Based Microtubes. *J. Am. Chem. Soc* **131**:8368 – 8369. doi: 10.1021/ja902684b.

[25]   Boulay, A.G., Cooper, G.J.T. & Cronin, L. (2012) Morphogenesis of polyoxometa-
       late cluster-based materials to microtubular network architectures. *Chem Commun.*
       48:5088 – 5090.
       doi: 10.1039/c2cc31194a.

[26]   Cooper, G.J.T., Boulay, A.G., Kitson, P.J., Ritchie, C., Richmond, C.J., Thiel, J.,
       Gabb, D., Eadie, R., Long, D.-L. & Cronin, L. (2011) Osmotically Driven Crystal
       Morphogenesis: A General Approach to the Fabrication of Micrometer-Scale Tubular
       Architectures Based on Polyoxometalates. *J. Am. Chem. Soc.* **133**:5947 – 5954.
       doi: 10.1021/ja111011j.

[27]   Cooper, G.J.T., Bowman, R.W., Magennis, E.P., Fernandez-Trillo, F., Alexander, C.,
       Padgett, M.J. & Cronin, L. (2012) Directed Assembly of Inorganic Polyoxometalate-
       based Micrometer-Scale Tubular Architectures by Using Optical Control. *Angew.
       Chem. Int. Ed.* **51**:12754 – 12758.
       doi: 10.1002/anie.201204405.

[28]   Symes, M.D., Kitson, P.J., Yan, J., Richmond, C.J., Cooper, G.J.T., Bowman, R.W.,
       Vilbrandt, T. & Cronin, L. (2012) Integrated 3D-printed reactionware for chemical
       synthesis and analysis. *Nature Chem.* **4**:349 – 354.
       doi: 10.1038/nchem.1313.

[29]   Kitson, P.J., Rosnes, M.H., Sans, V., Dragone, V. & Cronin, L. (2012) Configurable
       3D-Printed millifluidic and microfluidic 'lab on a chip' reactionware devices. *Lab
       chip* **12**:3267 – 3271.
       doi: 10.1039/c2lc40761b.

Beilstein-Institut

167

Beilstein Bozen Symposium on Molecular Engineering and Control
May 14th – 18th, 2012, Prien (Chiemsee), Germany

1 nm Thick Functional Carbon Nanomembrane (CNM): New Opportunities for Nanotechnology

Min Ai and Armin Gölzhäuser*

Physics of Supramolecular Systems and Surfaces, University of Bielefeld, Universitätsstraße 25, 33615 Bielefeld, Germany

E-Mail: *ag@uni-bielefeld.de

Received: 4th July 2013 / Published: 13th December 2013

Abstract

One nanometer thick, mechanically stable carbon nanomembranes (CNMs) are made by electron induced cross-linking of surface bound self-assembled monolayers (SAMs). The cross-linked SAMs are then released from the surface and can be placed onto solid materials or spanned over holes as free-standing membranes. Annealing at ~1000K transforms CNMs into graphene or graphenoids accompanied by a continuous change of mechanical stiffness and electrical resistance from insulating to conducting, which allows the tailoring of the CNM's electrical and mechanical properties. Recently, Janus membranes, i. e. CNMs functionalized by coupling different molecules to their top and bottom surfaces were built. Janus membranes have been built with functional polymers, proteins, and dyes, which demonstrates that Janus CNMs can act as platforms for two-dimensional chemistry. By combining different types of CNMs, hybrid nanolayers and biomimetic membranes can be built.

Introduction

Thin films are broadly and successfully integrated in many commercial products. They are used for surface lubrication, corrosion inhibition, or to avoid bio-fouling; other applications include barrier and separation membranes, light-emitting diodes, photo detectors, transistors, memory, and bio (chemical) sensors [1]. Most of these applications require a high thermal and mechanical stability while it is desirable that the films are as thin as possible. Commercially available thin films have thicknesses ranging from a few 100 nm to several µm. However, due to their extremely large surface-to-volume ratio, we expect that films with a much smaller thickness (< 100 nm) would be even more attractive for fundamental studies as well as for industry. Several ways to fabricate ultrathin films are currently explored: layer-by-layer (LbL) films of polyelectrolytes [2], spin-coating [3], interfacial polymeric membranes [4], Langmuir-Blodgett [5], chemical vapor deposition, as well as the exfoliation of bulk materials into graphene [6, 7] or transition metal dichalcogenides [8].

Other fabrication strategies are based on molecular self-assembly. Self-assembled monolayers (SAMs) of amphiphilic molecules have the thickness of a single molecule, ~1 nm. When aromatic SAMs are exposed to radiation, they form cross-links between neighboring molecules. The resulting molecular network can be released from the surface as a free-standing two-dimensional carbon nanomembrane (CNM). As amphiphilic molecules have two distinct functional groups, the resulting CNM also has two sides of distinct chemical functionality, and both sides can be chemically modified with dissimilar molecules, resulting in "Janus membranes". CNMs can also be converted into graphene by annealing in ultra-high vacuum. This diverse chemistry and broad functionality open new pathways for fundamental and applied research.

This article summarizes recent progress in the fabrication, characterization and application of CNMs. The main text is organized into two sections, the first emphasizing on the synthesis and preparation strategies that have been developed for achieving the 1 nm thick CNMs, and the analysis techniques that were used to characterize their physical and chemical properties. The second part is to discuss prospects of their applications in nanotechnology, e. g. being potentially integrated in lab-on-a-chip technology, electronics and micro-/nano-electro mechanical systems.

FABRICATION AND CHARACTERIZATION OF CNMs

1. Fabrication

The first step in the making of CNMs is the preparation of a well-defined self-assembled monolayer of aromatic amphiphilic molecules on a surface. A SAM can be obtained by adsorption from solution or by vapor phase deposition in vacuum. Depending on types of molecules and surfaces, SAMs with different degrees of order form. These SAMs are then irradiated by electrons [9] or UV-light [10], which starts a dehydrogenation and recombination mechanism that leads to a two-dimensional cross-linking of molecules into a CNM [11]. After cross-linking, CNM is detached from the original surface and exhibits a free-standing ultra-thin film with a thickness of the length of the SAM molecule (about 1 nm for a biphenyl molecule) [12]. Examples are presented in Table 1.

Table 1. CNM formation from molecules on appropriate substrates by electron beam and agents.

SAM	Anchor moiety	Substrate	Removing agent
Biphenylthiol or Nitrobiphenylthiol	-S	Au	I_2
Biphenylthiol or Nitrobiphenylthiol	-S	Cu	Ammonium persulfate + water
Biphenylcarboxylic acid	-COOH	Ag_2O	Ammonia solution (aq)
Biphenylhydroxamic acid	-CO-NH-OH	Fe/Ti	Acid/E-beam
Biphenylsilane or Hydroxybiphenyl	-OH or $-SiR_3$	Si	HF or KOH
Biphenylcarbonxylic acid or Nitrobiphenylcarboxylic acid	-COOH	ITO	Acid

2. Characterizing 1-nm-thick CNMs

Several analytical techniques for characterizing CNMs have been exploited. Among them optical microscopy, scanning electron microscopy (SEM), atomic force microscopy (AFM), helium ion microscopy (HIM), and transmission electron microscopy (TEM) provided an opportunity to study the morphology, optical and mechanical properties of CNM [13]. To elucidate the structure and composition of CNMs, spectroscopic techniques, such as infrared reflection absorption spectroscopy (IRRAS), X-ray and UV photoelectron spectroscopy (XPS, UPS), and Raman spectroscopy were utilized [9].

2.1 Imaging CNM

A 1 nm thick CNM is transparent in visible light. However, when it is placed onto a ~300 nm SiO_2 layer on a Si substrate, it can be detected with the naked eye, cf. Fig. 1 [13]. In Fig 1B, ribbons of CNMs on SiO_2/Si appear dark and blue shifted with respect to the substrate color in optical micrographs and two or three layers stacked on each other yield a higher contrast and blue shift, because of interference contrast, which results from light passing through layered structures with different dielectric properties, here the CNMs and the supporting SiO_2 on Si. Since interference contrast results in a clearly visible color change of the areas covered by CNMs and other thin nanofilms, it becomes a simple way of imaging such thin films.

CNMs were also imaged by helium ion microscopy (HIM). After being transferred onto metal grids, they became free-standing suspended on the holey support. Fig. 1C displays a free-standing CNM that spans over ~40 μm wide openings in metal grids, where finer structures, wrinkles, and defects are visualized. Scanning electron microscopy (SEM) can also be utilized to image CNMs by transferring them onto a perforated (2 μm holes) carbon foil and directly imaging them by SEM (Fig. 1C (right)). If the same sample is imaged by HIM, (Fig. 1C (left)), the helium ion image reveals a much higher contrast than the SEM image.

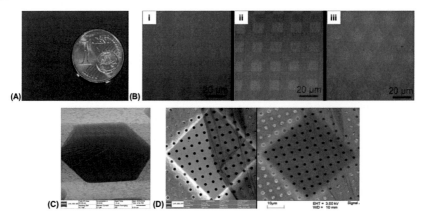

Figure 1. (A) Photograph of a cm^2-sized CNM (larger than one cent euro coin) on a SiO2/Si wafer. The CNM is visible due to Raleigh's interference contrast. **(B)** Optical micrograph of ~10 μm width CNM ribbons on a SiO2/Si wafer. i) A single layer. ii) Two layers transferred at a ~90° angle. iii) Three layers at an angle of ~60°. Each layer gives rise to a further blue shift. **(C)** HIM images of a CNM transferred onto a hexagonal pore (~40 μm in diameter). **(D)** Comparison of HIM and SEM images of a 1 nm thick CNM on a quantifoil TEM support.

2.2 Spectroscopic observation of the cross-linking

X-ray photoelectron spectroscopy was utilized to investigate the composition of SAMs, as well as radiation induced changes. Figure 2 displays C 1 s, N1 s, S 2 p, and O1 s spectra of 4-nitro-1,1'-biphenyl-4-thiol on a gold substrate, as recorded before and after irradiation by VUV under ultrahigh vacuum (UHV) conditions. The shift of the N1 s binding energy (BE) from 405.6 eV to 399.3 eV results from the chemical conversion of nitro groups to amino groups after irradiation with VUV. Characteristic changes also occur in the S 2 p and O1 s signals. A broadening of the S 2 p spectrum towards higher binding energies is attributed to a partial oxidation of the thiol species at the SAM/gold interface. The O1 s signal is significantly reduced after irradiation with VUV, which is explained by a loss of oxygen from nitro groups when they convert into amino groups. The thickness of SAM or CNM on the gold substrate can be determined from the attenuation of the Au4f$_{7/2}$ signal by the monolayer. The film thicknesses of SAMs (1.25 ± 0.8 nm) and CNMs (1.21 ± 0.9 nm) are obtained from these spectra [10].

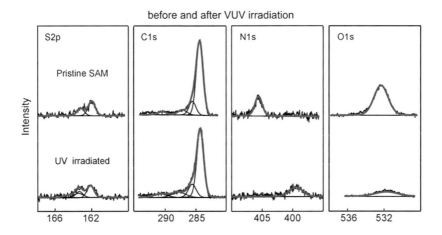

Figure 2. XP spectra of 4'-nitro-1,1'-biphenyl-4-thiol on gold before and after VUV (He I, 21.2 eV) irradiation. The generation of amino groups can be clearly recognized from the corresponding chemical shift from 405.5 eV to 399.2 eV [10].

2.3 Thermal stability

Thermal stability is important for the integration of CNMs in electronic devices. Turchanin, El-Desawy, and Gölzhäuser [14] found that cross-linked SAMs are thermally stable up to 1000 K. In their experiment, pristine SAMs and CNMs were heated on Au substrates in steps of 20 – 50 K from room temperature to 1000 K, and then studied by XPS [14]. In Fig. 3A, after heating to 470 K, the C 1 s signal of the pristine SAMs showed only 10% of its initial intensity, indicating SAM decomposition or desorption, but the C 1 s signal of CNMs appears at 90% intensity. Figure 3B shows a plot of the intensity of the C 1 s signal of pristine and electron irradiated BPT as a function of temperature and dose. The pristine SAMs desorbed at about 400 K, and the XPS intensity of CNM after irradiation at 25 mC/cm^2 electron dosage lead to 50% loss. Upon annealing to 1000 K, 80% of the CNM XPS intensity was still observed after 45 mC/cm^2 electron dosage irradiation. Lower electron dosage leads to the partially cross-linked SAM.

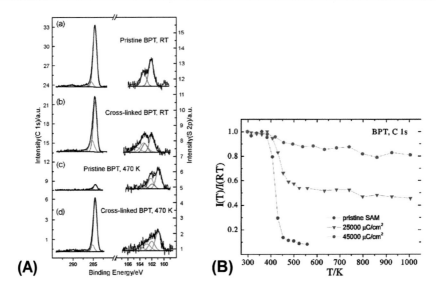

Figure 3. (A) XPS spectra of the C 1 s and S 2 p signals. (a) Pristine biphenythiol SAM (BPT) at room temperature. (b) Cross-linked BPT (45 mC/cm²) at RT. (c) Pristine BPT after annealing at 470 K. (d) Cross-linked BPT after annealing at 470K. **(B)** Temperature dependence of the C 1 s signal of BPT as a function of irradiation dose [14].

2.4 Electrical resistance

The performance of CNMs in electronic devices, of course, depends on their electrical conductivity. The sheet resistance of CNMs that have been annealed at temperatures between 800 K and 1200 K was determined by a two-point measurement in UHV scanning tunneling microscopy (STM) and by four-point in ambient conditions. CNMs suspended on a gold grid were directly contacted by a STM tip in UHV. The resistance was determined from the current between the tip and the surrounding frame (Fig. 4A). CNMs were also transferred onto non-conductive silicon oxide and were then measured with four-probes (Fig. 4bB) [12]. Figure 4C shows the sheet resistivity as a function of the annealing temperature. After annealing CNMs at ~800 K, we find a sheet resistivity of ~10^{-8} kΩsq^{-1}, upon rising the annealing temperature to ~1200 K, this comes down to ~100 kΩsq^{-1}. The reason for this unusual behavior is that upon annealing CNMs transform into graphene and thus become conductive, as will be described in more detail in section 2.7.

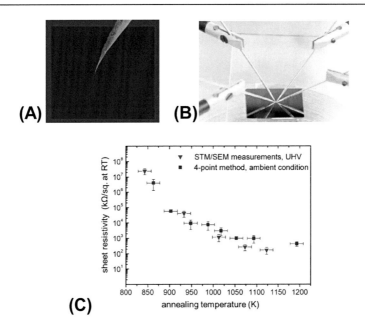

Figure 4. The resistivity of CNMs was determined by a two-point (STM) in UHV **(A)** and four-point in ambient conditions **(B)**. **(C)** The sheet resistivity corresponds to 10^{-8} kΩsq^{-1} after annealing at 800 K. Upon annealing to temperatures between 800 and 1200 K, the current/voltage curves are linear. Increasing the annealing temperature to 1200 K, the sheet resistivity drops to 100 kΩsq^{-1} [12].

Multilayers of CNMs after annealing show more conductivity than a single layer of CNMs, as in Fig. 5 [15]. Upon annealing to 1100 K, the sheet resistivity dropped further to 10.8, 30.3, and 37.2 kΩs q^{-1}, for five, four and three layers, respectively. This tunable electrical conductivity of CNMs from an insulator to a conductor by annealing temperatures or amounts of layers will be of value for CNMs to be utilized in micro- or nano-electronics.

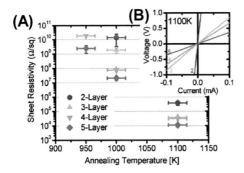

Figure 5. (A) RT sheet resistivity of BPT multilayers as a function of annealing temperature. The error bars represent the standard deviation of measurements at different spots on the samples. **(B)** RT linear current–voltage curves for 2-, 3-, 4-, and 5-layer samples after annealing at 1100 K [15].

2.5 Elastic properties

The elasticity of CNMs was investigated by Zhang, Beyer and Gölzhäuser [16] by using a bulge test method in an AFM for the determination of the Young's modulus and other quantities (Fig. 6). Gas pressure was applied to one side of a CNM that was freely suspended on a silicon substrate with a µm^2 opening. The pressure difference between the top and bottom side of the membrane resulted in a CNM deflection that was recorded by line-scanning the tip of an AFM (Fig. 6A) or by monitoring the deflection of a fixed AFM tip (central-point method, Fig. 6B). The central-point method has the advantage of reducing data acquisition time and lowering the probability that the CNM ruptures during the experiment. Young's modulus and internal stress have been calculated from the obtained pressure-deflection relationship. Maximum deflections several µm under at pressures up to 2 kPa were measured. The ultimate tensile strength reached 440 – 720 MPa with elongation to break at values between 3 – 4%, and the Young's modulus was obtained in the range of 6 to 12 GPa. The Young modulus depends on the precursors of CNMs or the electron doses for the cross-linking CNMs. Below 20 mC/cm^2, no free-standing CNMs were formed, whereas between 30 mC/cm^2 and 50 mC/cm^2, CNMs were formed. With further irradiation, the Young's moduli remained constant, even when the CNM was irradiated with much higher doses of up to 80 mC/cm^2. This result was in accordance with the thermal stability experiments of CNMs, which indicated almost complete cross-linking at an electron dose of ~45 mC/cm^2 [16].

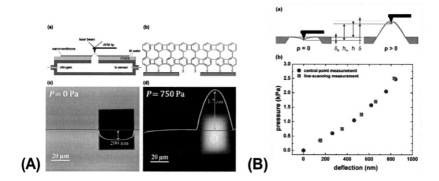

Figure 6. (A) (a) Schematic diagram of an AFM bulge test; (b) CNM suspended over an opening on Si substrate. (c) AFM image of a non-pressurized CNM and a line profile with a downward deformation of 200 nm; (d) CNM under pressure of 750 Pa with an upward 1.7 µm deflection. **(B)** (a) Schematic of the central-point method; (b) Comparison of the line-scanning and the central-point method [16].

2.6 Chemical modification of CNM surfaces

The chemical modification of CNMs primarily depends on the functionality of the precursor molecules. When CNMs made from 4'-nitro-1,1'-biphenyl-4-thiol molecules are electron irradiated, the hydrogen released during the cross-linking is reducing the terminal nitro groups to amino groups. These can be further functionalized with other molecules[17, 18], polymers [19 – 21] and proteins [22], as is shown in the schematic representation in Fig. 7.

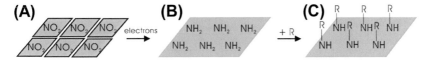

Figure 7. Scheme of electron induced chemical lithography: **(A)** 4'-nitro-1,1'-biphenyl-4-thiol monolayer. **(B)** An electron beam converts the terminal nitro groups of a 4'-nitro-1,1'-biphenyl-4-thiol monolayer to amino groups while the underlying aromatic layer is cross-linked. **(C)** The cross-linked amino biphenyl thiol region is used for the selective coupling of molecules R (small molecules, or proteins, or polymers).

An example are polymer brushes that can be grown onto CNMs by surface-initiated polymerization (SIP) [20, 21], or self-initiated surface photopolymerization and photografting (SIPGP) [22]. In Fig. 8, poly(4-vinyl pyridine) (P4VP) brushes were prepared on CNMs by SIPGP. In different solvents and depending on the pH values the morphology of the polymer brushes changes. In aqueous solution at pH = 7, the P4VP brushes are non-transparent on glass slides (Fig. 8B). This results from strong buckling due to collapsed polymer chains with low solubility. In ethanol or aqueous solutions at pH = 2.5, the P4VP brushes appear transparent on glass and flat on gold substrate (Fig. 8A and 8C), as ethanol is a very good solvent for P4VP and at low pH values the polymers tend to dissolve in aqueous solution due to the protonated pyridine group. This visible response of the CNM bound polymer brush to solvent stimulus may be useful for actuators, sensors, or displays, providing opportunities to adjust a surface to an environmental stimulus.

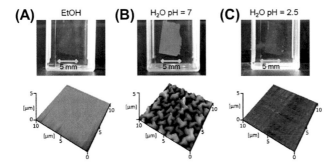

Figure 8. Photographs (above) and AFM measurements (below) of P4VP carpets in **(A)** ethanol, **(B)** water at pH 7, and **(C)** water at pH 2.5 [22].

CNMs can also be decorated with distinct functional groups on both sides. One side of the CNM is occupied with the amino groups as aforementioned, while the other bears sulfur residues after separation from Au. Both sides can react with other groups, and "Janus" CNMs that utilize this particularity have been constructed by Zheng *et al.* [24], who immobilized electron donors tetramethylrhodamine (TMR) to the top and electron acceptors (ATTO647N) to the bottom side of the same CNM. The synthesis processes are schematically displayed in Fig. 9. Figure 10 shows fluorescence micrographs and SEM images of CNMs functionalized on its amino side with TMR and on the thiol side with ATTO647N. The fluorescence resonance energy transfer (FRET) efficiency from the electron donor to the electron acceptor through the short spacer (~1 nm thick CNM) was determined to be effectively 100%, while the distance is much shorter than the Förster radius. This demonstration of "Janus" CNMs could lead to a platform for two-dimensional chemistry [24].

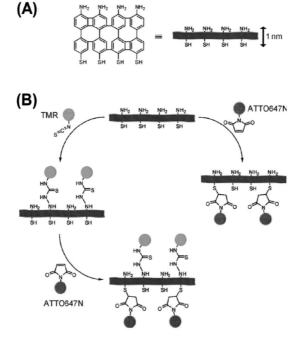

Figure 9. (A) A nanomembrane made from cross-linked biphenyl self-assembled monolayers. The 1 nm thick membrane has amino and thiol functional groups on its upper and lower sides, respectively. **(B)** Functionalization with fluorescent dyes of the upper and lower sides of the nanomembranes. The amino side is functionalized with TMR, the thiol side with ATTO647N.

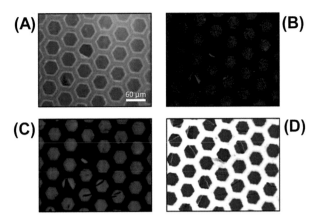

Figure 10. (A, B) Fluorescence and **(C)** FRET images, and **(D)** the corresponding SEM image of a nanomembrane functionalized on its upper and lower sides with TMR and ATTO647N, respectively. The FRET image has been recorded by exciting at the donor wavelength and recording the fluorescence emission in the acceptor channel. All micrographs show the same area of the nanomembrane [23].

2.7 Transforming CNMs into graphene

CNMs transform into graphene when they are heated in ultrahigh vacuum to temperatures up to 1200 K (pyrolysis). This has been demonstrated by Turchanin *et al.* [12, 15] by investigating the pyrolyzed CNM with the help of Raman spectroscopy, conductivity measurements (also see 2.4.), and transmission electron microscopy. The Raman spectrum of untreated CNMs does not show any peaks between 1000 and 2000 cm^{-1}. Upon annealing, the G-peak at 1605 cm^{-1}, that is characteristic for graphene, appears. Parallel, a rise of up to 5 orders of magnitude in the electrical conductivity is observed when the annealing temperature increases from room temperature to 1200 K, see figure 4. This is due to the formation of nanosized graphene crystallites in annealed CNMs. Furthermore, ordered domains comprised of individual atoms in annealed CNMs were observed by HRTEM (Fig. 11). Utilizing this bottom-up route from CNMs, graphene can be manufactured with well-defined thicknesses and tunable electrical properties on various metal and insulator substrates. This unique resistance behavior allows to tune CNMs from insulator to conductor.

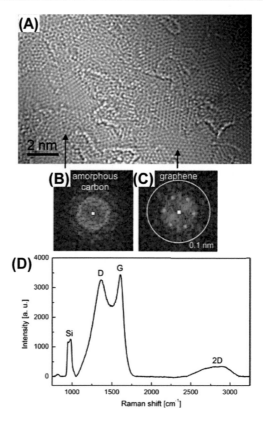

Figure 11. (A-C) High-resolution transmission electron micrograph (80 kV) of graphene, $T_{an} = 1200$ K fabricated from CNMs. The graphene sheet was transferred onto a TEM grid. **(A)** Unprocessed image; **(B and C)** local Fourier transforms from 3×3 nm^2 areas indicated by the arrows. The presence of diffraction spots in panel **(C)** verifies the presence of crystalline graphene areas in panel **(A)**. **(D)** Raman spectrum (excitation wavelength, 532 nm) of the same sample transferred to a SiO2/Si surface.

OPPORTUNITIES FOR NANOTECHNOLOGY

Interest in CNMs is motivated by their unique properties, the tunable electrical conductivity, stiffness and surface chemistry. A variety of applications for CNMs have been proposed, and here we discuss CNMs as nanotemplates, nanopores and in biochips.

Nanotemplates are low dimensional substrates having artificially or naturally patterned structures. They can be created by non-covalent interactions, for example, hydrogen bonding, metal-organic coordination and charge transfer, van der Waals interactions [24]. The patterned structures can provide a controlled shape and composition; nanotemplates have attracted much interest in molecular engineering and recognition. Patterned SAMs can be simply fabricated by electron exposures [25]. Such patterns which contain irradiated and non-irradiated regions are useful templates for material deposition on the micro- and nanoscale.

Nanopores have the ability to interact with atoms, ions and molecules not only at the surfaces, but throughout the bulk. They are broadly found in nature, e.g. in zeolite minerals or biomembranes. In recent years artificial nanoporous materials were fabricated to be utilized for catalysis, gas separation, water purification and biomolecular separation [26]. Two-dimensional nanoporous materials, i.e. sieves, have a perspective in bioseparation or proton-exchange in fuel cell when being integrated in microfluidic or membrane systems.

Biochips, as used in microfluidic or "lab-on-chip" devices, reduce laboratory procedures and enable portable detection of agents. In the past decades, cell biotechnology for cell culture and secretion [27], protein biochips [28], DNA biochips[29] have been a significant expansion in research. The key to the success of microfluidic-based biotechnology is to immobilize biomolecules onto surfaces with high density. For integration of 2D CNMs in biochips, the immobilization of biomaterials is indispensable. Biomolecules such as proteins and cells have been reported to be immobilized on the chemically tailored CNMs or physically adsorbed on CNMs [10, 30].

1. Micro- and nanotemplates

Micro- and nanotemplates can be used to fabricate metallic nanostructures on electrodes. An electrochemical surface template can be generated with structured SAMs. The SAMs are patterned by electron exposures through stencil masks with holes down to $1\,\mu m$ and the subsequent electrodeposition of copper on SAM patterned gold electrodes has been demonstrated. In non-irradiated biphenyl SAMs, the copper peak in the cyclic voltammogram was observed at 0.08 V from the $CuSO_4/H_2SO_4$ electrolyte solution. For electron irradiated SAMs, the deposition peak was shifted to more negative potentials ($\Delta E = 0.15$ V). The electrodeposition of copper nanostructures was thus selectively generated on SAM templates at a deposition potential of +0.05V, where the irradiated biphenyl perfectly inhibited the copper deposition, as shown in Fig. 12 [31, 32].

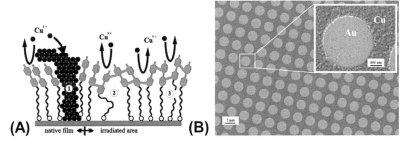

Figure 12. (A) Schematic illustration of the electrochemical behavior of a BP12 SAM. For the native film copper deposition is expected to be defect mediated (1). Electron irradiation links the biphenyl moieties which can then mask intrinsic defects (2) or radiation induced damage in the alkane spacer (3). **(B)** SEM-micrograph of Cu structures deposited on a BPT SAM template patterned by proximity printing (300 eV, $64000\,\mu C/cm^2$). Deposition parameters: 10 mM $CuSO_4$, +0.05 V (vs. NHE), 120 s [31, 32].

2. Micro- and nanopores

To drill holes in a 1-nm-thick film CNMs, fast and low-cost methods were demonstrated. Photolithography can be used to define the size and shape of pores that are then etched away by UV/ozone, cf. Fig. 13 [33]. In a more advanced method periodic nanopores are fabricated by extreme UV interference lithography (Fig. 14). The size of the nanopores can be flexibly tuned by choosing proper EUV masks, with a lower limit of currently ~20 nm [34].

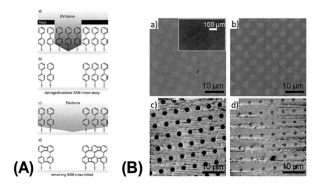

Figure 13. (A) (a) A SAM of BPT is exposed to UV/ozone through a mask. (b) The damaged/oxidized SAM in the exposed regions is rinsed away and (c) exposed to electrons. (d) The remaining SAM is cross-linked. **(B)** (a) SEM image of a BPT SAM after UV/ozone patterning through a TEM grid with 3 μm circular holes. (b) SEM after cross-linking (a) with an electron dose of 50 mC/cm^2. (c) LFM of a UV/ozone patterned BPT-SAM. (d) LFM of (c) after cross-linking [33].

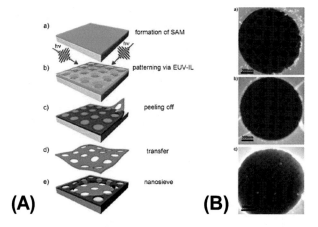

Figure 14. (A) Fabrication scheme of nanosieve membranes. (a) A SAM of NBPT is formed on a gold substrate. b) Generation of amino-terminated crosslinked areas (nanosheets) in the SAM via EUV-IL. c) Spin-coating of a polymer film and peeling off from the gold substrate. d) Transfer of the nanosheet with the hardened polymer. e) Formation of a 1-nm-thin freestanding nanosieve by dissolving the polymer on a new holey substrate.

(B) TEM characterization of nanosieve membranes. Dark-field mode images of ~1-nm-thin nanosieves with a) 138 ± 17-nm holes (interference grating periodicities of 225×200 nm^2; irradiation dose of 70 J cm^{-2}), b) 100 ± 15-nm holes (interference grating periodicities of 225×200 nm^2; irradiation dose of 120 J cm^{-2}), and c) 31 ± 6-nm holes (interference grating periodicities of 100×90 nm^2; irradiation dose of 50 J cm^{-2}) [34].

3. Biochips

3.1 Chemically tailored CNMs

To demonstrate CNMs as protein chips Turchanin, Tinazli and co-workers used the barrel-shaped 20S proteasome complex from the archaea Thermoplasma acidophilum as a biological model system periodically binding on CNMs [22]. The macromolecular proteins were engineered on CNMs via complexation between histidine-tagged proteins and transition metal ions, Ni (II). The Nickel-nitrotriacetic acid (NTA) and histidine-tagged chelator system is a powerful and universal tool for the one-step isolation and purification of gene products [35]. For the application in protein arrays and biosensors, long-term stability is required to overcome the limitation of metal leaching and rapid protein dissociation [36]. Therefore, the authors introduced cumulated NTA clusters on CNMs with multivalent interaction sites to load and regenerate His6-tags proteins. This concept of periodically binding proteins on CNMs shed light on immobilizing and then detecting single proteins as a biochip, as shown in Fig. 15.

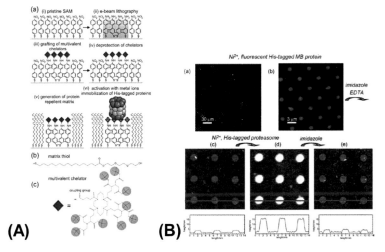

Figure 15. (A) Schematic representation of the protein chip assembly, (i–v); Controlled immobilization of His-tagged protein complexes (here: proteasome) in end-on orientation (vi). b) Protein repellent EG3-OH matrix thiols. c) Multivalent tris-NTA chelator with protected carboxyl functionality. **(B)** Controlled immobilization of different proteins on the chip surface. a, b) Confocal laser scanning microscopy images of specifically immobilized Atto565-labeled maltose-binding proteins (His10-MBP). c) Topographic AFM image of the same chip after the regeneration with imidazole. d) In situ AFM scan of a microarray of immobilized His6-proteasome complexes. e) The same surface after regeneration. AFM images are $14 \times 14 \, \mu m^2$, the z-scale corresponds to 20 nm [22].

3.2 Physically immobilized on CNMs

Non-covalent interaction (van der Waals and electrostatic forces) between biomolecules and CNMs provides an alternative and simple way to investigate the biocompatibility of CNMs. Rhinow *et al.* have prepared purple membranes (PMs) from *Halobacterium salinarum*, a 2-D crystalline monolayer of bacteriorhodopsin and lipids as a test specimen on CNMs [37]. Characterization of PM on CNMs by cryogenic high-resolution transmission electron microscopy (cryo-EM) (Fig. 16 B) reveals that trehalose embedded PM samples have been imaged on CNMs, and the amplitudes as well as phases can be recovered up to high resolution ranges. The biocompatibility of CNMs to PMs is further confirmed with tapping-mode AFM in a buffer solution. The 2D crystal lattice of PMs on CNMs was visualized, and the force spectroscopy obtained by AFM exhibited non-folding peaks, due to the extracellular side of PM's physical adsorption on CNMs [37].

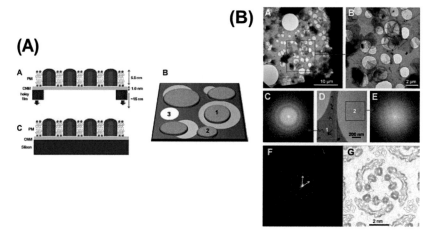

Figure 16. (A) Schematic representation of PMs on CNMs. (A) PMs on freestanding CNMs. PMs and CNMs are drawn on scale with respect to their thickness. (B) (1) PMs sticking to freestanding CNMs, (2) PMs supported by CNMs plus thick carbon support, and (3) holey carbon without CNMs. (C) PMs on silicon-supported CNMs.

(B) (A) TEM image of trehalose-embedded PMs on CNMs at room temperature. A holey carbon film supports the CNM. (B) Enlarged view of the central area of (A) (red square). (C) FFT of area 1 of (D). Thon rings are visible due to phase contrast of the thick holey carbon support. (D) TEM image of a sample-free CNM area, which spans two adjacent holes. (E) FFT of free-hanging CNM (area 2 of (D)). Thon rings are absent. (F) Fourier transform of an image of a single trehalose-embedded PM on cCNM at liquid nitrogen temperature. White arrows indicate the unit cell vectors corresponding to the 62.4 Å lattice. The red arrowhead points to the (4, 3) reflection at 8.9 Å resolution. (G) A projection map of PMs on CNMs at 4 Å resolution as calculated from eight averaged images in a defocus range of 500 – 1000 nm [37].

CONCLUSIONS AND OUTLOOK

This article provides an overview on recent trends in CNM research. The fabrication of CNMs from various self-assembled small aromatic molecules of a variety of substrates has been reviewed. The exploration of the properties of these two-dimensional films was revisited, among them the tuning of its chemical, mechanical and electronic function, and its conversion to graphene after pyrolysis. Janus CNM in which the top of the CNM is decorated with electron donors and while its bottom with electron acceptors, and their used as "rulers" to determine the length (generally 1 nm ~ 8 nm) of a molecular spacer between donor and acceptor incorporated by biomolecules is discussed. When both sides of the CNM are decorated with site-specific biomolecules, CNMs can be utilized to mimic biological membranes. There is a range of further applications in nanoscience and nanotechnology, for instance, micro- or nanotemplates, micro- or nanopores and biochips. The tunable conductivity and surface chemistry will permit CNMs to be integrated in lab-on-a-chip devices for liquid or gas separation, electronics, MEMS/NEMS devices in the near future. Opportunities and challenges co-exist, for example, in developing new routes for preparing CNMs of macroscopic size, and understanding the detailed nanostructures of CNMs. These unsolved fundamental questions together with potential applications will provide compelling motivation for CNM research in the coming years.

ACKNOWLEDGEMENTS

The described work was made possible by financial support from the Volkswagenstiftung, the Deutsche Forschungsgemeinschaft, and the German Bundesministerium für Bildung und Forschung (BMBF).

REFERENCES

[1]   Huang, G., Mei, Y. (2012) *Adv. Mater.* **24**:2517 – 2546.
        doi: 10.1002/adma. 201200574.

[2]   Decher, G., (1997) *Science* **277**:1232 – 1237.
        doi: 10.1126/science.277.5330.1232.

[3]   Vendamme, R., Onoue, S.-Y., Nakao, A., Kunitake, T. (2006) *Nature Materials* **5**:494 – 501.
        doi: 10.1038/nmat1655.

[4]   Louie, J. S., Pinnau, I., Reinhard, M. (2008) *J. Membr. Sci.* **325**:793 – 800.
        doi: 10.1016/j.memsci.2008.09.006.

[5]   Nardin, C., Winterhalter, M., Meier, W. (2000) *Langmuir* **16**:7708 – 7712.
        doi: 10.1021/la000204t.

[6]   Li, X., Cai, W., An, J. (2009) *Science* **324**:1312 – 1314.
doi: 10.1126/science.1171245.

[7]   Novoselov, K. S., Geim, A. K., Morozov, S. V., Jiang, D., Zhang, Y., Dubonos, S. V.,
Grigorieva, I. V., Firsov, A. A. (2004) *Science* **306**:666 – 669.
doi: 10.1126/science.1102896.

[8]   Chhowalla, M., Shin, H. S., Eda, G., Li, L.-J., Loh, K. P., Zhang, H. (2013) *Nature Chemistry* **5**:263 – 275.
doi: 10.1038/NCHEM. 1589.

[9]   Eck, W., Küller, A., Grunze, M., Völkel, B., Gölzhäuser, A. (2005) *Adv. Mater.* **17**:2583 – 2587.
doi: 10.1002/adma.200500900.

[10]   Turchanin, A., Schnietz, M., El-Desawy, M., Solak, H. H., David, C., Gölzhäuser, A. (2007) *Small* **3**:2114 – 2119.
doi: 10.1002/smll.200700516.

[11]   Turchanin, A., Käfer, D., El-Desawy, M., Wöll, Ch., Witte, G., Gölzhäuser, A. (2009) *Langmuir* **25**:7342 – 7352.
doi: 10.1021/la803538z.

[12]   Turchanin, A., Beyer, A., Nottbohm, C. T., Zhang, X., Stosch, R., Sologubenko, A., Mayer, J., Hinze, P., Weimann, T., Gölzhäuser, A. (2009) *Adv. Mater.* **21**:1233 – 1237.
doi: 10.1002/adma.200803078.

[13]   Gölzhäuser, A., Wöll, Ch. (2010) *ChemPhysChem.* **11**:3201 – 3213.
doi: 10.1002/cphc.201000488.

[14]   Turchanin, A., El-Desawy, M., Gölzhäuser, A. (2007) *Appl. Phy. Lett.* **90**:053102.
doi: 10.1063/1.2437091.

[15]   Nottbohm, C. T., Turchanin, A., Beyer, A., Stosch, R., Gölzhäuser, A. (2011) *Small* **7**:874 – 883.
doi: 10.1002/smll.201001993.

[16]   Zhang, X., Beyer, A., Gölzhäuser, A. (2011) *Beilstein J. Nanotechnol.* **2**:826 – 833.
doi: 10.3762/bjnano.292.

[17]   Gölzhäuser, A., Eck, W., Geyer, W., Stadler, V., Weimann, T., Hinze, P., Grunze, M. (2001) *Adv. Mater.* **13**:806 – 809.
doi: 10.1002/1521-4095(200106)13:11%3C803::AID-ADMA806%3E3.0.CO;2-W.

[18]   Nottbohm, C. T., Sopher, R., Heilemann, M., Sauer, M., Gölzhäuser A. (2010) *J. Bio. Tec.* **149**:267 – 271.
doi: 10.1016/j.jbiotec.201001018.

[19]   Schmelmer, U., Jordan, R., Geyer, W., Eck, W., Gölzhäuser, A., Grunze, M., Ulman, A. (2003) *Angew. Chem. Int. Ed.* **42**:559 – 563.
doi: 10.1002/anie.200390161.

[20]   Schmelmer, U., Paul, A., Küller, A., Steenackers, M., Ulman, A., Grunze, M., Gölzhäuser, A., Jordan, R. (2007) *Small* **3**:459 – 465.
doi: 10.1002/smll.200600528.

[21]   Amin, I., Steenackers, M., Zhang, N., Beyer, A., Zhang, X., Pirzer, T., Hugel, T., Jordan, R., Gölzhäuser, A. (2010) *Small* **6**:1623 – 1630.
doi: 10.1002/smll.201000448.

[22]   Turchanin, A., Tinazli, A., El-Desawy, M., Großmann, H., Schnietz, M., Solak, H. H., Tampé, R., Gölzhäuser A. (2008) *Adv. Mater.* **20**:471 – 477.
doi: 10.1002/adma.200702189.

[23]   Zheng, Z., Nottbohm, C. T., Turchanin, A., Muzik, H., Beyer, A., Heilemann, M., Sauer, M., Gölzhäuser A. (2010) *Angew. Chem. Int. Ed.* **49**:8493 – 8497.
doi: 10.1002/anie.201004053.

[24]   Cicoira, F., Santato, C., Rosei, F. (2008) *Top Curr. Chem.* **285**:203 – 267.
doi: 10.1007/128_2008_2">10.1007/128_2008_2.

[25]   Gölzhäuser, A., Geyer, W., Stadler, V., Eck, W., Grunze, M., Edinger, K., Weimann, T., Hinze, P. (2000) *J. Vac. Sci. Technol. B* **18**:3414 – 3418.
doi: 10.1116/1.1319711.

[26]   Davis, M. E. (2002) *Nature* **417**:813 – 821.
doi: 10.1038/nature00785.

[27]   Yang, M., Yao, J., Duan, Y. (2013) *Analyst* **138**:72 – 86.
doi: 10.1039/c2an35744e.

[28]   Rusmini, F., Zhong, Z., Feijen, J. (2007) *Biomacromolecules* **8**:1775 – 1789.
doi: 10.1021/bm061197b.

[29]   Dutse, S. W., Yusof, N. A. (2011) *Sensors* **11**:5754 – 5768.
doi: 10.3390/s110605754.

[30]   Biebricher, A., Paul, A., Tinnefeld, P., Gölzhäuser, A., Sauer, M. (2004) *Journal of Biotechnology* **112**:97 – 107.
doi: 10.1016/j.jbiotec.2004.03.019.

[31] Kaltenpoth, G., Völkel, B., Nottbohm, C.T., Gölzhäuser, A., Buck, M. (2002) *J. Vac. Sci. Techno. B* **20**:2734 – 2738.
doi: 10.1116/1. 1523026.

[32] Völkel, B., Kaltenpoth, G., Handrea, M., Sahre, M., Nottbohm, C. T., Küller, A., Paul, A., Kautek, W., Eck, W., Gölzhäuser, A. (2005) *Surface Science* **597**:32 – 41.
doi: 10.1016/j.susc.2004. 08. 046.

[33] Nottbohm, C.T., Wiegmann, S., Beyer, A., Gölzhäuser A. (2010) *Phys. Chem. Chem. Phys.* **12**:4324 – 4328.
doi: 10.1039/B923863H.

[34] Schnietz, M., Turchanin, A., Nottbohm, C. T., Beyer, A., Solak, H. H., Hinze, P. T., Weimann, Gölzhäuser A. (2009) *Small* **5**:2651 – 2655.
doi: 10.1002/smll.200901283.

[35] Hochuli, E., Bannwarth, W., Döbeli, H., Gentz, R. & Stüber, D. (1988) *Bio-Techniques* **6**:1321 – 1325.
doi: 10.1038/nbt1188-1321.

[36] Nieba, L., NiebaAxamnn, S. E., Persson, A., Hamalainen, M., Edebratt, F., Hansson, A., Lidholm, J., Magnusson, K., Karlsson, A. F., Pluckthun, A. (1997) *Anal. Biochem.* **252**:217 – 228.
doi: 10.1006/abio.1997.2326.

[37] Rhinow, D., Vonck, J., Schranz, M., Beyer, A., Gölzhäuser, A., Hampp, N. (2010) *Phys. Chem. Chem. Phys.***12**:4345 – 4350.
doi: 10.1039/b923756a.

BIOGRAPHIES

Nediljko Budisa

studied Chemistry, Biology, Molecular Biology and Biophysics at the University of Zagreb (Croatia). He made his PhD work in the laboratory of Robert Huber at the Max-Planck-Institute of Biochemistry in Martinsried and defended his PhD thesis (summa cum laude) in 1997 at the Technical University, Munich. During PhD work he initiated an independent project on protein design and engineering by an expanded amino acid repertoire.

These works were further elaborated and extended during his Postdoctoral work (1997 – 2000) in Martinsried in the laboratories of Robert Huber and Luis Moroder and laid the ground for his habilitation work (finished 2005) at the Technical University in Munich. He started an independent research Group 'Molecular Biotechnology' at the Max-Planck-Institute of Biochemistry after receiving the prestigious BioFuture Award of the German Ministry for Research and Education in 2004. After vocation in 2008, he holds the Chair of Biocatalysis at TU Berlin from the May 2010.

Nediljko Budisa's research at the interface of chemistry and biology is mainly concerned with the possibilities to generate novel generic classes of amino acids by using synthetic chemistry, bio-computing and target-engineered metabolic circuits. Such novel amino acids should enable an expansion of the genetic code in the context of a reprogrammed protein translation. The final goal is to create designer molecules and cells carrying out novel properties optimized for user-defined environments or to engineer a genetic code such that proteome-wide substitutions/insertions of synthetic amino acids can yield a synthetic life with new chemical possibilities.

Timothy Clark

studied chemistry at the University of Kent at Canterbury, UK and obtained his Ph.D. in physical chemistry at the Queen's University Belfast, Northern Ireland. After a postdoctoral stay in Princeton, he joined the Friedrich-Alexander Universität Erlangen-Nürnberg, where he is now Director of the Computer-Chemie-Centrum and also of the Centre for Molecular Design at the University of Portsmouth, UK. His research involves computational investigations of phenomena ranging from protein-DNA interactions and signal-transduction in biological systems to modeling the physical, electronic and spectroscopic properties of advanced molecular materials and nanosystems. He is a member of the board of the Excellence Cluster "Engineering of Advanced Materials" and Editor in Chief of the Journal of Molecular Modeling. He has published two books and more than 330 scientific papers.

Lee Cronin's

research aims to do fundamental, creative, and profound studies in the field of inorganic chemistry, specifically the self-assembly and self-organization of inorganic molecules and the engineering of complex systems leading to the emergence of system-level behaviours. Contributions include the development of new techniques to control the assembly of nanoscale molecular metal oxide clusters, some of the largest non-biological molecules known, the development of new cryospray and variable temperature mass spectrometry techniques for the elucidation of reaction mechanism and the observation of highly reactive intermediates as well as the discovery of emergent nano/micro structures such as tubules, membranes and inorganic cells.

Projects in the group include developing minimal inorganic systems capable of evolution, engineering materials with complex and emergent behaviours, as well as the development of new reaction formats for complex and novel chemistry e. g. flow systems and 3D-printing.

Armin Gölzhäuser

studied physics at the University of Heidelberg and at Arizona State University. In 1993 he received his Ph.D. from the University of Heidelberg (with Prof. Michael Grunze). From 1993 – 1996 he was a Feodor Lynen Fellow at the University of Illinois at Urbana-Champaign with Prof. Gert Ehrlich. He then returned to Heidelberg where he received his habilitation in 2001.

In 2003 he became an associate professor of physical chemistry at the University of Marburg. Later in the same year, he became a full professor of physics at the University of Bielefeld, where he stayed since then.

Armin Gölzhäuser's research is focused on the fabrication and characterization of functional materials, and he utilizes radiation induced chemistry to transform molecules into nanostructures. His laboratory operates modern analytical tools, among them electron microscopes, photoelectron spectrometers, scanning probes and the first helium ion microscope at a German university.

Armin Gölzhäuser has a strong interest in the technological application of nanostructures. In 2011, he founded CNM Technologies GmbH, a start-up company dealing with applications of carbon nanomembranes.

He is the speaker of the Bielefeld Institute for Biophysics and Nanoscience (BINAS). He organized scientific conferences and served in committees of national and international scientific societies and he is an associate editor of the Beilstein Journal of Nanotechnology.

Martin G. Hicks

is a member of the board of management of the Beilstein-Institut. He received an honours degree in chemistry from Keele University in 1979. There, he also obtained his PhD in 1983 studying synthetic and theoretical approaches to the photochemistry of pyridotropones under the supervision of Gurnos Jones. He then went to the University of Wuppertal as a post-doctoral fellow, where he carried out research with Walter Thiel on semi-empirical quantum chemical methods. In 1985, Martin joined the computer department of the Beilstein-Institut where he worked on the Beilstein Database project. His subsequent activities involved the development of cheminformatics tools and products in the areas of substructure searching and reaction databases.

Thereafter, he took on various roles for the Beilstein-Institut, including managing director-ships of subsidiary companies and was head of the funding department 2000 – 2007. He joined the board of management in 2002; his current interests and responsibilities range from organization of Beilstein Symposia with the aim of furthering interdisciplinary communica-tion between chemistry and neighbouring scientific areas, to the publishing of the Beilstein Open Access journals – Beilstein Journal of Organic Chemistry and Beilstein Journal of Nanotechnology – and production of scientific videos for Beilstein.TV.

Michael Huth

2002 –	University professor at Physikalisches Institut, Johann Wolf-gang Goethe University Frankfurt am Main
2001 – 2002	Assistant professor at Institut für Physik, Johannes Gutenberg University Mainz
2001	Habilitation in experimental physics
1998 – 2001	Postdoc at Institut für Physik, Johannes Gutenberg University Mainz
1997 – 1998	DFG research scholarship at Physics Department and Materials Research Laboratory, University of Illinois at Urbana-Cham-paign; Group of Prof. Dr. Colin Peter Flynn
1995 – 1997	Postdoc at Institut für Physik, Johannes Gutenberg University Mainz
1995	PhD in Physics at Technical University of Darmstadt
1990	Diploma in Physics at Technical University of Darmstadt

- Member of advisory board for the Institute for Microtrechnologies (IMTech), University of Applied Sciences RheinMain, Rüsselsheim

- Member of advisory board of Beilstein-Institut, Frankfurt, Foundation for furthering the chemical and related sciences

- Member of advisory board of NanoScale Systems GmbH, Darmstadt

- Co-founder of Nanoscale Systems GmbH, Darmstadt

- Spokesperson for NanoNetwork Hessen (NNH) for area Frankfurt (Goethe University, University of Applied Sciences)

Carsten Kettner

studied biology at the University of Bonn and obtained his diploma at the University of Göttingen. In 1996 he joined the group of Dr. Adam Bertl at the University of Karlsruhe and successfully narrowed the gap between the biochemical and genetic properties, and the biophysical comprehension of the vacuolar proton-translocating ATP-hydrolase using the patch clamp techniques. He was awarded his Ph.D for this work in 1999. As a post-doctoral student he continued both the studies on the biophysical properties of the pump and investigated the kinetics and regulation of the dominant plasma membrane potassium channel (TOK1).

In 2000 he moved to the Beilstein-Institut to represent the biological section of the funding department. Here, he is responsible (a) for the organization of the Beilstein symposia (ESCEC, Glyco-Bioinformatics and Beilstein Bozen Symposium) and the publication of the proceedings of the symposia and (b) for the administration and project management of funded research projects such as the Beilstein Endowed Chairs (since 2002), the collaborative research centre NanoBiC (since 2009) and the Beilstein Scholarship program (since 2011). In 2007 he started a correspondence course at the Studiengemeinschaft Darmstadt (a certified service provider) where he was awarded his certificate of competence as project manager for his studies and thesis.

Since 2004 he coordinates the work of the STRENDA commission and promotes along with the commissioners the proposed standards of reporting enzyme data (www.strenda.org). These reporting standards have been adopted by more than 28 biochemical journals for their instructions for authors and are subject for the development of an electronic data capturing tool. With this background, in 2011, Carsten was appointed to co-ordinate another standardization project (MIRAGE) which is concerned with the uniform reporting and representation of glycochemistry data in publications and databases.

Andreas Kirschning

studied chemistry at the University of Hamburg and Southampton University (UK). In Hamburg, he joined the group of Prof. E. Schaumann and received his PhD in 1989 working in the field of organosilicon chemistry. After a postdoctoral stay at the University of Washington (Seattle, USA) with Prof. H. G. Floss, supported by a Feodor-Lynen scholarship of the Alexander-von Humboldt foundation, he started his independent research at the Clausthal University of Technology in 1991, where he finished his habilitation in 1996. In 2000 he moved to the Leibniz University Hannover and became a director of the institute of organic chemistry. He is one of the editors of RÖMPP online, *Natural Products Reports*, *Journal of Flow Chemistry* and *Beilstein Journal of Organic Chemistry*. His research interests cover structure elucidation as well as the total synthesis and mutasynthesis of natural products, target elucidation of natural products, biomedical biopolymers, and finally synthetic enabling technologies (solid-phase assisted synthesis, minituarized flow reactors, inductive heating).

Robert Macfarlane

Education

Sept. 2007 –	Northwestern University, PhD Candidate in Chemistry
Sept. 2005 – May 2006	Yale University, MS in Inorganic Chemistry
Sept. 2000 – May 2004	Willamette University, B.A. in Biochemistry

Professional Experience

2009 – 2012	Bionanomaterials Group Leader, Mirkin group. Supervised and coordinated the research endeavours of ~25 graduate students and post-doctoral researchers in the areas of chemistry, biochemistry, materials science, and biology.
2008 – 2012	Graduate4 Student Mentor, Mirkin group Directly mentored 5 graduate students in their research process, including research project development, laboratory techniques, data analysis and scientific communication skills.
2008	Teaching assistant, General Chemistry, Northwestern University Chem. Dept.
2007	Teaching assistant, Thermodynamics, Northwester University Chem. Dept.
2006	Teaching assistant, Occasional Lecturer, General Chemistry, Yale University Chem. Dept.
2002 – 2004	Tutor, Organic Chemistry, Willamette University

<u>Honors and Awards (selection)</u>

2011	Materials Research Society Gold Graduate Student Award
2011	International Precious Metals Institute Sabin Corp. Graduate Student Award
2010	International Institute for Nanotechnology Outstanding Researcher Award
2009	Ryan Fellowship, Northwestern University

Paul S. Weiss

is director of the California NanoSystems Institute, Fred Kavli Chair in NanoSystems Sciences, and distinguished professor of chemistry & biochemistry and of materials science & engineering at the University of California, Los Angeles. He received his S.B. and S.M. degrees in chemistry from MIT in 1980 and his Ph.D. in chemistry from the University of California at Berkeley in 1986. He was a postdoctoral member of technical staff at Bell Laboratories from 1986 – 1988 and a visiting scientist at IBM Almaden Research Center from 1988 – 1989.

Before coming to UCLA in 2009, he was a distinguished professor of chemistry and physics at the Pennsylvania State University, where he began his academic career as an assistant professor in 1989. His interdisciplinary research group includes chemists, physicists, biologists, materials scientists, electrical and mechanical engineers, and computer scientists. Their work focuses on the atomic-scale chemical, physical, optical, mechanical and electronic properties of surfaces and supramolecular assemblies. He and his students have developed new techniques to expand the applicability and chemical specificity of scanning probe microscopies. They have applied these and other tools to the study of catalysis, self- and directed assembly, physical models of biological systems, and molecular and nanoscale devices. They work to advance nanofabrication down to ever smaller scales and greater chemical specificity in order to connect, to operate, and to test functional molecular assemblies, and to connect these to the biological and chemical worlds. Two current major themes in his laboratory are cooperativity in functional molecules and single-molecule biological structural and functional measurements.

Weiss has been awarded a NSF Presidential Young Investigator Award (1991 – 1996), the B. F. Goodrich Collegiate Inventors Award (1994), an Alfred P. Sloan Foundation Fellowship (1995 – 1997), the ACS Nobel Laureate Signature Award for Graduate Education in Chemistry (1996), a John Simon Guggenheim Memorial Foundation Fellowship (1997), and a NSF Creativity Award (1997 – 1999), among others. He was elected a Fellow of: the American Association for the Advancement of Science (2000), the American Physical Society (2002), the American Vacuum Society (2007), and the American Chemical Society (2010), and an Honorary Fellow of the Chinese Chemical Society (2010). He is the founding

Editor-in-Chief of ACS Nano (2007 –). At ACS Nano, he won the Association of American Publishers, Professional Scholarly Publishing PROSE Award for 2008, Best New Journal in Science, Technology, and Medicine, and ISI's Rising Star Award a record ten times.

Dave A. Winkler

studied chemistry, chemical engineering, and physics at the Monash and RMIT Universities in Melbourne. He completed a PhD in microwave spectroscopy and radioastronomy at Monash University in 1980 as a General Motors postgraduate fellow. He then worked as a tutor and senior research fellow, helping establish the computer-aided drug design group at the Victorian College of Pharmacy with Prof. Peter Andrews. He joined the Commonwealth Scientific and Industrial Research Organization (CSIRO) in Melbourne as a in 1985 where he worked on the design of biologically active small molecules as drugs and agrochemicals. He was awarded Australian Academy of Science traveling fellowships to Toshio Fujita's lab in Kyoto in 1988, and to Graham Richards' research group in Oxford in 1997, and a Newton Turner Fellowship for exceptional senior scientists in 2010.

His research with CSIRO has a common theme of the application of computational methods to understanding and designing novel small molecules with biological activity. During the last decade he has also worked with a diverse complex systems science group in Australia and internationally. This aimed to understand how complexity concepts such as emergence, criticality, self-organization and self-assembly can contribute to the modelling, understanding and design of complex chemical and biological systems. In the past seven years, his research has evolved to encompass biomaterials and regenerative medicine. He currently leads two research projects; one involved in understanding regulatory processes in stem cells, and how they trigger fate decisions from a complexity perspective; the other is using more traditional molecular modelling methods to design small molecule mimics of cytokines that influence stem cell fate. These projects involved collaboration with the Australian Stem Cell Centre and international stem cell groups.

Dave is a Fellow and past Chairman of the Board of the Royal Australian Chemical Institute, an Adjunct Professor at Monash University, past President of Asian Federation for Medicinal Chemistry, an enduring organizer of Pacifichem, and on the Board of Science and Technology Australia. He has published over 200 scientific papers, reports, and patents, and co-edited four multi-author books. He is on the Editorial boards of *ChemMedChem* and *BMC Biophysics*.

Biographies

Index of Authors

Index

197

Index   Beilstein Bozen Symposium on Molecular Engineering and Control, May 14th – 18th, 2012, Prien (Chiemsee), Germany

198

Beilstein Bozen Symposium on Molecular Engineering and Control, May 14ᵗʰ – 18ᵗʰ, 2012, Prien (Chiemsee), Germany **Index**

199

Index   Beilstein Bozen Symposium on Molecular Engineering and Control, May 14th – 18th, 2012, Prien (Chiemsee), Germany